Visual Telephony

Visual Telephony

Edward A. Daly
Kathleen J. Hansell

Artech House
Boston • London

Library of Congress Cataloging-in-Publication Data
Daly, Edward A.
  Visual telephony / Edward A. Daly, Kathleen J. Hansell.
    p.   cm. — (Artech House telecommunications library)
  Includes bibliographical references and index.
  ISBN 1-58053-023-0 (alk. paper)
  1. Video telephone industry. I. Hansell, Kathleen Jane. II. Title. III. Series.
HD9697.T452D35   1999
651.7'3—dc21   99-10794
                                                               CIP

British Library Cataloguing in Publication Data
Daly, Edward A.
  Visual telephony. – (Artech House telecommunications library)
  1. Video telephone
  I. Title II. Hansell, Kathleen J.
  621.3'88'5

  ISBN 1-58053-023-0

Cover design by Lynda Fishbourne

© 1999 ARTECH HOUSE, INC.
685 Canton Street
Norwood, MA 02062

International Standard Book Number: 1-58053-023-0
Library of Congress Catalog Card Number: 99-10794

10 9 8 7 6 5 4 3 2 1

This book is dedicated
to my wife and best friend, Janet,
with gratitude
for her inspiration and encouragement.

Ed Daly

Contents

Part 1

Getting Started With Visual Telephony

V ISUAL TELEPHONY IS not a single subject area but rather a combination of many distinct fields—audio, video, data, multimedia, transmission, compression, networking, and even psychology. It is no wonder that new users feel bewildered and betrayed when they find out that the so-called "simple" solution the vendor tried to sell wasn't such a simple solution after all.

The best way to get started with visual telephony is to begin to understand what it is, in general terms, and to learn where to go for information and assistance. That's what Part 1 is all about. Chapter 1 defines visual telephony and describes how it is used in a variety of applications. Chapter 2 suggests sources of expertise and warns of pitfalls to avoid as you begin your journey toward implementing a system.

1

Visual Telephony: Applications and Benefits

V ISUAL TELEPHONY IS the transmission of visual images electronically. This text does not attempt to cover all of the uses of visual
telephony, such as cable television, which fits the broad definition
of visual telephony. Rather, in this book we are concerned with a
subset of the uses of visual telephony that has the following
characteristics:

- *Multidirectional.* Unlike broadcast television, where viewers watch
  video but are themselves never seen by others, the visual telephony applications discussed in this book are those that make it
  possible for all participants to both see and be seen. Although
  not all designs allow everyone to view everyone else continuously, the designs do make it possible for any one participant
  to see any other participant as the session progresses.

- *Simultaneous.* The transmission and reception of images take place synchronously (at the same time).

- *Motion.* The images move with normal, or nearly normal, motion.

1.1   Applications

Visual telephony has many uses. Here is a short description of the most common applications.

1.1.1   Videoconferencing

A videoconference uses visual telephony to create an electronic meeting in which people participate from separate locations rather than gathering together in one place. This is the dominant use of visual telephony. However, a significant number of visual telephony systems have been installed to serve other purposes.

1.1.2   Distance education

Among the earliest adopters of visual telephony were educators, who saw great possibilities for the technology. Distance education can bring learning to students who might not otherwise have such opportunities. Students in rural communities can study advanced subjects with experts. Adults in the workplace can learn while they are on the job. Students all over the globe can enroll in college and university programs of study.

1.1.3   Telemedicine

Remote diagnostics, teleradiology, and remote case management are among the medical uses of visual telephony that are commonly categorized as *telemedicine*. Other medical applications may or may not be categorized as telemedicine. For example, when administrators in several hospitals meet electronically to share solutions to common problems, the application is traditional videoconferencing. However, when doctors consult with other physicians on a diagnosis, and unique tools are required to do so, the application is better classified as telemedicine. Likewise, the training of medical staff can be viewed as distance education; but when course work requires equipment that is singularly applicable to medical training, the application is better described as telemedicine.

1.1.4 Judicial applications

The technology offers a number of benefits to the judicial system. The most common application is video arraignments. People arrested without a warrant must be formally charged with a crime, a process known as an *arraignment*. Although arraignment proceedings take only a few minutes, transporting the person to be arraigned can cost thousands of dollars. Many courts have installed video links with detention centers for video arraignments.

In large cities, a crime victim must travel to the police station near the crime scene to meet with officers investigating the case. When an arrest is made, the victim must come to a police station to identify the suspect. The victim may also be asked to meet with representatives of the district attorney's office to review the case and prepare for trial. The victim may spend hours traveling to and from police stations and DA offices for these appointments. By means of visual telephony systems installed at police stations and prosecutors' offices, the victim can be spared some of the inconvenience of these trips.

Parole hearings are another application of visual telephony. Rather than traveling among prisons, members of the parole board can meet at a central site and connect electronically to each prison. Travel costs are avoided, the members are less stressed, and the cost of additional security while the board members are at the prison is eliminated.

Potential witnesses can give pretrial testimony, called depositions, over visual telephony links. In some instances, witnesses have been allowed to testify during the trial using visual telephony.

1.1.5 Sales and marketing applications

New uses for visual telephony are appearing in point-of-sale systems. Kiosks linked to experts can promote a company's products in a store.

In shopping centers and supermarkets, automatic teller machines (ATMs), enhanced with video links, provide full banking services to customers who are not conveniently served by a bank's major branch offices.

Visual telephony is also useful in market research. Focus groups can be interactively connected with marketing executives, who can observe group reactions and discuss products directly with group members.

1.1.6   Other applications

These are the current applications of visual telephony. Others, such as teledining, have been discussed but not implemented. As the technology becomes familiar to more people, there will no doubt be many more applications. One thing is certain—visual telephony is here to stay.

1.2   Why visual telephony?

In every application category, many good reasons exist for implementing visual telephony. One educational institution may invest in distance education to increase revenues by offering courses to students in another country. Another may be using distance education to cut costs by more fully utilizing its teaching talent. Yet another might be using the technology to enrich courses that are currently being taught.

Every organization that invests in visual telephony has some specific objective, or objectives, in mind. The most frequently cited objective for videoconferencing is *the reduction of travel costs.* Although there are no hard data to support this observation, it appears that most acquisitions are justified entirely, or in large part, by the anticipated reduction in travel-related costs. Yet after the acquisition, other benefits are typically found. Some years ago, AT&T undertook a survey of existing users to determine just how effective videoconferencing was in reducing travel costs. The outcome was surprising. Travel costs were avoided, but survey respondents ranked cost avoidance low on their lists of videoconferencing benefits. Of more importance to them were a number of other benefits:

- *Faster decision making.* Because a videoconference can be established with little more difficulty than an in-house meeting, people separated by miles or even oceans can come together and share ideas and information when the need arises. There is no need to delay until all participants can clear days from their calendars for travel.

■ *Better decisions.* There is no additional cost to include more meeting attendees. The conference can include all of the people who are involved in a project. Information sources, not originally considered, can be brought into the meeting as the need arises. People can join a videoconference as easily as they can an in-house meeting. If travel were required, this might not be possible, even if the organization were willing to incur the cost.

■ *Increased productivity.* Valuable employees are not wasting time traveling to and from airports, queuing for check-in, and waiting on the tarmac for take-off clearance. If a meeting takes 2 hours, the participants will expend only 2 hours.

■ *Travel costs avoided.* This is the most easily quantified benefit. A business trip requires the traveler to either pay for conveyance to the airport or pay for parking. At the destination, a rental car or a taxi may be required. Some trips require an overnight hotel stay. Nearly all trips call for the company to pay for meals. Airfare is a significant cost, and business travelers can seldom take advantage of discount rates that require advance payment or weekend stays.

■ *Greater employee morale.* Some business travel is enjoyable. However, frequent trips, particularly to remote locations or second-tier cities, can be a drain on employee morale and a strain on employees' families. *Videoconferencing cannot eliminate the need for travel.* It can, however, reduce travel requirements significantly.

■ *Increased employee safety.* Some corporations require that key employees travel separately. This reflects the fact that travel involves some risk. A videoconference eliminates this concern.

■ *Improved security.* Many people work on a business trip. In the cramped quarters of an aircraft, delicate information may be compromised if documents or laptop computer screens are seen by other travelers or if business discussions among coworkers are overheard.

1.2.1 Functional areas that can benefit from visual telephony

Specific benefits of visual telephony identified by current users are shown in Table 1.1. It is by no means complete. New uses of visual telephony appear constantly. Each new use provides the innovator with benefits previously not considered.

It was only a few years ago that people thought of visual telephony as science fiction. Today, most major organizations have a network of systems. Visual telephony is an established communications tool. Prices have declined to the point that even small organizations can now afford to implement visual telephony in one or more applications. *Any organization that depends on the interaction of people who are not under the same roof should be using visual telephony.*

Table 1.1
Benefits of Visual Telephony

Type of Organization	Functional Area	Specific Benefits
Corporation	Finance	More accurate budgeting
		Better cost control
	Sales	Better management of sales resources
		Lower training costs
		Improved account management
		More executive contact with key customers
		Better finished goods inventory control
	Operations	Improved quality
		Better work-in-progress inventory control
		Fewer product returns
	Marketing	Shorter time to market with new products
		Products better suited to market
		More thorough market analysis
		More focus groups
		Entry into new markets
	Personnel	Lower recruiting costs
		Less turnover

Table 1.1
Benefits of Visual Telephony

Type of Organization	Functional Area	Specific Benefits
Educational institution	Course delivery	Better course access
		More consistent presentation
		More diversified classes
		Better use of teaching resources
	Administration	New revenue sources
	Personnel	Opportunities for faculty consulting
		Better coordination of multicampus departments
	Placement	Greater access to potential employers
		Access to more recruiters
Medical facility	Remote diagnostics	Lower patient transport costs
		Earlier intervention
		Expert opinions
		Higher occupancy rates at local hospitals
		Improved outcomes
	Training	Lower in-service training costs
		Quicker delivery
	Teleradiology	Lower staffing costs
		Improved staff morale
	Case management	Lower personnel costs
		Convenience for patients
		Fewer admissions
Judicial facility	Initial proceedings	Reduced security costs
		Better utilization of police
		More efficient use of court space
	Depositions	Reduced costs
		Faster discovery
	Testimony	Access to important witnesses
		Faster trials
		Protection of juvenile witnesses
	Witness participation	Improved witness cooperation
		Better utilization of police
		Faster verification
	Parole hearings	More frequent hearings
		Less cost
		More security
Entertainment industry	Auditions	Lower cost
		Greater diversity
		Quicker decisions
	Directing	Remote access
		Lower production costs

2

Sources of Expertise

THE FIRST FRUSTRATION potential buyers encounter is the lack of a well-defined marketplace for visual telephony. Some first-time buyers have a difficult time locating anyone with industry knowledge. Others encounter a number of market participants willing to provide a small portion of a visual telephony system, but with limited knowledge of the rest of the industry.

2.1 Main elements

A visual telephony installation can be considered to have four main elements: the environment, the system components, the codec, and the network.

2.1.1 Environment

Whether you need a conference room for videoconferencing, a classroom for distance education, or an examining room for telemedicine, all visual telephony applications require a suitable environment. That environment should be appropriate for both viewing a video image and sending one, and for both listening to and being heard at the remote end. Not all sites are perfectly suited for visual telephony. Some need more modification than others. Typical modifications include lighting, acoustics, and electrical work. (Chapter 13 provides recommendations for site selection and preparation.)

2.1.2 System components

For interactive applications of visual telephony, every installation must have at least one video camera, one video monitor, one speaker, and one microphone. Few systems contain this minimal configuration. Most are more complex. (See Chapters 8 through 12 for detailed discussions of system components.)

2.1.3 Codec

The element that has allowed visual telephony to enter the mainstream of communications is the digital video compression coder/decoder, or *codec*. By digitally compressing video signals, the codec allows affordable telephone connections to be used for videoconferencing. A codec is not included in every system, but without codecs and lower cost transmission connections, fewer visual telephony applications would be cost justified. (Chapter 16 is devoted to a discussion of this very important element.)

2.1.4 Network

Every system must have a means of conveying signals among visual telephony sites. Visual telephony is much more complicated than voice telephony, and a wide range of options exists. (Network options are discussed in detail in Chapters 14, 15, 17, and 18.)

2.2 Potential providers

Finding someone with knowledge and experience in audio/visual design and integration, video compression, data communications, and telephone networking is not an easy task. Here are some possible sources of expertise.

2.2.1 System integrators

Some companies specialize in integrating audio/visual components into complete systems. A few can analyze a buyer's specific requirements and available facilities, prepare environmental and systems designs, and perform the integration of components into a functional and appropriate system. Typically, system integrators like to provide custom-designed solutions to meet a customer's specific requirements. But they should be able to do more than that. Because most organizations find that the correct videoconferencing system for one location is not the most appropriate for others, a variety of system types is often required. If the integrator recommends expensive custom-built systems for every situation, beware. The integrator should be able to supply a range of ready-made systems as well as custom-built systems as required.

2.2.2 Audio/visual dealers

Most audio/visual dealers lack the expertise required to design and implement an interactive visual telephony system. Using these companies can be very risky. Some companies will research visual telephony and install a system, learning on the job. The resulting system often painfully illustrates the inexperience of the designer. In addition, these systems tend to be expensive, because the provider wants to recover research expense from one client.

The most significant drawback to using audio/visual dealers is their geographic limitation. Most videoconferencing applications require a national or international network of installations. While the leading system integrators can install and maintain systems nationally, most local audio/visual companies cannot service a geographically dispersed network.

For applications that serve local areas (for example, distance education within a community), a local A/V company may be the ideal provider. A few have had the opportunity to learn about a specific application and have become quite good at providing these unique services. Most A/V companies, however, lack sufficient understanding of visual telephony, a shortcoming that can be overcome by partnering with a more experienced market participant.

In general, when dealing with a system integrator or A/V dealer, look for competence in all the skill sets described earlier. Ask for references. Be sure the organization is listening to your needs. And be sure to ask for a detailed after-sale support plan.

2.2.3   Codec manufacturers

The manufacturers of video codecs compete vigorously for market share, producing both codecs and standardized systems that incorporate their codecs. The systems are sold by the manufacturers directly to end-users and through dealers, distributors, and other retail outlets. Systems sold by codec manufacturers represent well over half of all visual telephony sales in the United States.

Codec manufacturers from around the world participate in the North American market. Some have the ability to install and maintain the systems they sell nationally; others have a global capability. A few have neither. Some codec manufacturers can ably assist in connecting their systems to the telephone network; others cannot.

Like system integrators, codec manufacturers should be able to provide references. They should also be able to provide a detailed support plan.

2.2.4   Telephone companies

The global telecommunications industry pioneered the development of visual telephony. In most countries, the national telephone company (known as the *Post, Telegraph, and Telephone* authority, or *PTT*) was the first to bring visual telephony to market. In the United States, a number of pioneering organizations began in the early 1980s to promote visual telephony for videoconferencing applications. AT&T and the telephone companies that made up the Bell System introduced videoconferencing under the brand name PicturePhone Meeting Service™. A creative satellite transmission company, Satellite Business Systems (SBS), promoted its own design and service offering. SBS was then acquired by MCI. Another innovator, Isacomm, pioneered videoconferencing systems and public videoconference rooms. It was acquired by Sprint.

Today's dominant long-distance telephone carriers can all claim a role in the development of visual telephony, and they continue to be actively involved in it.

Lucent, formerly AT&T's equipment sales organization, and MCI act as retailers for video telephony equipment manufactured by others. After-sales support may be provided by the carrier or the manufacturer, a subject the buyer should carefully investigate.

Since divestiture, the Regional Bell Operating Companies (or RBOCs), which include BellSouth, Bell Atlantic, and U S West, have made some effort to sell visual telephony. In general, they act as sales agents for codec manufacturers. They solicit orders for the manufacturer in return for a commission. The sales contract is between the manufacturer and the customer. The buyer must look to the manufacturer for after-sales support. Once again, the buyer should ask for references and a support plan.

2.3   Industry associations

Several organizations exist that have members involved in visual telephony or its specific applications. These organizations can provide useful information to those new to visual telephony. Here are several.

International Teleconferencing Association

The International Teleconferencing Association (ITCA) is an international nonprofit association that promotes the growth and development of visual telephony and other forms of collaborative communications. The organization provides a host of programs and services for its members, including annual conferences, and serves as a resource for information on the collaborative industry. Contact information for this association follows:

International Teleconferencing Association
100 Four Falls Corporate Center, Suite 105
West Conshohocken, Pennsylvania 19428
U.S.A.

Phone: +1 610 941 2020
Fax: +1 610 941 2015

E-mail: staff@itca.org
Web site: http://www.itca.org

United States Distance Learning Association

The United States Distance Learning Association (USDLA) advocates and promotes distance learning via visual telephony and other electronic technology in education, business, health care, and government. It represents the distance learning community before government policy and regulatory bodies and is a sponsor of major distance learning conferences. Contact information for this association follows:

United States Distance Learning Association
1240 Central Boulevard, Suite A
Brentwood, California 94513
U.S.A.

Phone: +1 800 275 5162 or +1 925 513 4253
Fax: +1 925 513 4255
E-mail: charles@usdla.org
Web site: http://www.usdla.org

American Telemedicine Association

The American Telemedicine Association (ATA) is a nonprofit association promoting greater access to medical care via telecommunications technology. It offers peer interaction, participation in public policy task forces, educational materials, and conferences. Contact information for this association follows:

American Telemedicine Association
901 15th Street, N.W.
Suite 230
Washington, DC 20005
U.S.A.

Phone: +1 202 408 0677
Fax: +1 202 408 1134
E-mail: atmeda@atmeda.org
Web site: http://www.atmeda.org

Telemedicine Research Center

The Telemedicine Information Exchange (TIE), a web site maintained by the Telemedicine Research Center, is useful to those interested in telemedicine. This site offers databases, articles, information on sources of funding, an on-line forum, and links to other resources. Contact information for this center follows:

Telemedicine Research Center
2121 SW Broadway, Suite 130
Portland, Oregon 97201
U.S.A.

Phone: +1 503 221 1620
Fax: +1 503 223 7581
E-mail: trc@telemed.org
Web site: http://tie.telemed.org

Association of Telemedicine Service Providers

The Association of Telemedicine Service Providers (ATSP) is an organization for health care providers that are involved in telemedicine services. It offers educational materials, information, member services, and an annual conference. Contact information for this association follows:

Association of Telemedicine Service Providers
7276 S.W. Beaverton-Hillsdale Highway, Suite 400
Portland, Oregon 97225
U.S.A.

Phone: +1 503 222 2406
Fax: +1 503 223 7581
E-mail: info@atsp.org
Web site: http://www.atsp.org

2.4 Standards development

Visual telephony cannot be adequately discussed without emphasizing the role that interoperability standards have played in making a visual telephony industry possible.

The companies that pioneered the development of visual telephony products worked independently. Their resulting products were not compatible. As a result, the earliest adapters of videoconferencing found themselves stranded on a technology island, with no way to bridge to other users, who were stranded on similar islands. Products did not interoperate. Networks could not be connected. National telephone companies could not connect internationally.

Through the International Telecommunications Union (ITU), a United Nations chartered organization based in Geneva, global communications industry participants have developed standards that today ensure international connectivity. ITU standard H.320 is the set of standards for visual telephony. The portion of H.320 that deals with video is H.261. Other standards deal with audio, network connectivity, graphics, and other elements associated with visual telephony. These standards are described in detail in Chapter 16. Contact information for the ITU follows:

International Telecommunications Union (ITU)
Place des Nations
CH-1211 Geneva 20
Switzerland

Phone: +41 22 730 5111
Fax: +41 22 733 7256
E-mail: itumail@itu.int
Web site: http://www.itu.ch

Part 2

Applications and Typical Installations

VISUAL TELEPHONY CAME on the scene in the late 1970s and early 1980s as a way to have an "electronic meeting" with people participating from various locations in a visual and audio exchange. It was called *videoconferencing* to differentiate it from the more generic term *teleconferencing,* which came to be descriptive for both videoconferences and audio-only sessions called *audioconferencing.*

Even as primitive as videoconferencing was in those days, with giant equipment boxes, clumsy controls, and jerky motion images, people could see the promise of the rapidly advancing technology for a host of applications beyond mere meetings.

Part 2 discusses today's applications—videoconferencing (Chapter 3), distance education (Chapter 4), telemedicine (Chapter 5), legal applications (Chapter 6), and applications in sales and marketing (Chapter 7). In addition, each chapter describes typical systems to support unique uses within the application area.

3

Videoconferencing

VIDEOCONFERENCING IS THE term used to describe an electronic meeting in which video and audio are used to exchange ideas. The meeting may involve corporations, educational institutions, hospitals or health care organizations, or other entities. The application may be a project review, a planning session, a problem-solving meeting, an interview, a staff meeting, a budget review, or some other kind of meeting. As long as it involves people meeting via video, the activity is considered *videoconferencing*.

3.1 Basic equipment

Videoconferencing installations vary widely. This chapter discusses the various types of installations and how they differ. First, however, let's discuss aspects that all videoconferencing sites have in common.

- *Video display*. The most apparent feature of a videoconferencing installation is the video display, which shows an image of the distant conference participant(s). This display must be large enough for all local participants to see without difficulty. In some cases, the display will be a desktop computer monitor for one person. At the other extreme, there may be a 6-foot-long projection screen for hundreds of people. Consumer television sets or multiple large-screen professional monitors might also display video for videoconferencing applications.

- *Camera.* A video camera is situated very near the video display(s). It is used to capture the image of the local conference participant(s). There may be an additional camera for whiteboard or podium presentations or for other special requirements. In most cases, there is also a camera for imaging printed material and solid objects.

- *Microphones and speakers.* There will be one or more microphones. These are typically low profile designs that are inconspicuous. There will also be one or more audio speakers.

- *Other equipment.* Ancillary equipment, such as a videocassette recorder (VCR) or facsimile machine, may be included in the installation.

- *Controller.* A controller allows the users to establish a video connection and to control the system during a conference. Some systems are controlled with a mouse and on-screen icons as well as a keyboard to enter dialing information. Other systems have either an infrared handset, a mechanical control box, or a touch-sensitive screen. Some systems have more than one type of controller, and the user can select the controller with which he or she is most comfortable.

3.2   Installations for a wide range of applications

A wide range of videoconferencing installations is available to any organization, from high-end custom-built rooms to low-end portable units.

3.2.1 Custom-built rooms

Custom-built rooms are typically well-appointed conference facilities (Figure 3.1). The visual telephony equipment is integrated into the room. Room layout, furniture, lighting, wall covering, and decor are all selected to enhance videoconferences. The cost of custom-built rooms varies with the complexity and appointments. Prices start at slightly under $100,000; the amount that can be spent is unlimited. However, most installations seem to cost about $250,000.

Most custom facilities are added to existing work locations. The modification of the space and the installation of equipment must be

Figure 3.1 A custom-built videoconferencing facility. This videoconferencing room utilizes a common partition for the conference and equipment rooms. An opening was framed in the partition to receive a custom-made cabinet of equipment that allows access from the equipment room. The room features a custom-made table. (*Source:* Systematics General Corporation, Herndon, Virginia.)

coordinated to avoid conflicting with normal work activities. This significantly contributes to the cost. When a custom room is included in new construction or as part of an extensive renovation, the cost is less. In fact, in this case the cost is about the same as a rollabout system.

3.2.2   Rollabout group systems

Videoconferencing equipment can be factory installed in transportable cabinets. Because the amount of on-site work is reduced to between 8 and 16 hours, labor costs are significantly lower. Roughly 70% of videoconferencing installations today are rollabout systems.

Rollabout systems can be divided into two categories depending on the group sizes for which they are suitable.

- *Large group systems.* These systems include one or two monitors that are between 25 and 37 inches measured diagonally (Figure 3.2). The size of the monitors allows groups of three or more to view the incoming video comfortably. The cameras have lenses that can capture three or four seated participants in a single view. The system may have multiple cameras or the capability

Figure 3.2   A large group rollabout system. This system features dual large-screen monitors. (*Source:* Tandberg, Herndon, Virginia.)

to move the camera remotely to focus on all of the main seating locations. There are multiple microphones. Large group systems support a wide range of optional equipment. The typical large group system costs between $25,000 and $75,000.

■ *Small group systems.* The obvious difference between small group systems and large group systems is size. Small group systems typically have only one monitor, and it may be smaller than the monitors used in large group systems (Figure 3.3). However, the most significant difference is the minimal capabilities supported by the system. Small group systems usually offer only one microphone and a single camera. They are designed for up to four people participating in a videoconference with limited audio/visual support material. Most manufacturers offer the ability to increase a small group system's capabilities. However, as the system is enhanced, its price quickly crosses over the purchase price of a large group system similarly equipped. The

Figure 3.3 A small group rollabout system. Systems like this are suitable for offices and small conference rooms. (*Source:* Sony Videoconferencing, Park Ridge, New Jersey.)

typical cost of a small group system ranges from $10,000 to $30,000.

3.2.3 Desktop video systems

One of the most exciting, and most misunderstood, segments of the videoconferencing market is the desktop arena. The term *desktop system* can be misleading, because it can refer to both stand-alone video systems and computer-based systems. *Here the term will mean noncomputer small units that reside on a desktop to support only videoconferencing.*

Some desktop video systems are designed to provide up to two people with face-to-face video and audio connectivity. For some applications, these are an economical alternative to small group systems. Other desktop units are practical only for use by one person.

A desktop unit for up to two people includes a small monitor (9 or 10 inches), a fixed camera, and a built-in microphone and speaker (Figure 3.4). Few optional items can be added. These units cost between $4,500 and $8,000.

A personal desktop unit for a single user will have a smaller screen (5 inches or less), a built-in camera, speakerphone, and a handset for private conversations (see Figure 3.5). These products are called video telephones or videophones. They cost between $2,000 and $8,000.

Desktop video products often include a data port for computer-to-computer communications. Units that comply with international standards can be interactively connected to computer-based systems through these data ports.

3.2.4 Computer-based videoconferencing systems

These products are hardware and software enhancements to computer systems that permit the user to videoconference while working on a computer.

Some very inexpensive packages are available that allow a standard desktop computer such as a PC or Mac to send and receive images and audio over standard telephone lines and the Internet. These packages include an inexpensive camera that attaches to, or rests on, the computer monitor, and a microphone. These products do not include any videoconference boards. They rely instead on software that uses

Figure 3.4 This system is designed to serve one or several users in an office environment. (*Source:* PictureTel Corporation, Andover, Massachusetts.)

the microprocessor of the computer to process the incoming and outgoing video. These products operate over standard telephone lines. The video and audio quality are not very good, which severely limits applications. They might be of interest to computer hobbyists but are not very useful for business applications. Computer processors will need to become five times more powerful than they currently are if software is to provide interactive video at a quality acceptable for business applications.

For this reason, *the discussion of computer-based desktop units that follows is limited to systems that include videoconference boards,* a design feature that provides much better quality than current software-only products.

Figure 3.5 This desktop system occupies a small space on the user's desk and serves a single user or several participants. (*Source:* Tandberg, Herndon, Virginia.)

Such systems include a camera, videoconference and sound boards, speakers, and microphone that are added to multimedia PCs or Macs or to more powerful computer systems such as the Sun Microsystems and Digital Equipment Corporation workstations. An important distinction is that these systems use digital telephone connections. These products provide good video quality and are therefore well suited for many serious applications of visual telephony.

In most computer-based videoconferencing packages, the camera has a fixed focal length that will focus on a person sitting within 3 feet of the camera and monitor. An upgrade to the camera may allow it to be zoomed and focused on people farther than 3 feet from the camera and monitor. When more people are involved, a display larger than the 13- to 15-inch computer monitor is required. For these applications, some manufacturers include a 27-inch television monitor as a system option.

The two types of computer-based videoconferencing systems that use videoconference boards are (1) those designed to operate over local-area networks (LANs) and (2) those that operate over digital services provided by local and long-distance telephone carriers.

■ *Computer-based systems operating over LANs.* This type is designed to transmit digital images over an in-house data network, known as a LAN (Figure 3.6). Computers linked via a LAN can share files, and when the visual telephony hardware and software are added, users can see each other and talk with each other while they interactively work on text, spreadsheets, designs, and diagrams. The linking of computers via a LAN allows people to join participative groups without leaving their offices. Typically four or five people, each at his or her own computer, can be connected. Because LANs are designed for in-house connectivity,

Figure 3.6   A computer-based system. The addition of a codec card set transforms a typical computer to a high-quality conferencing system that can connect to the Internet or a corporate LAN server. (*Source:* VTEL Corporation, Austin, Texas.)

most LAN videoconferencing products include a "gateway" feature that allows LAN users to go outside the LAN and access standard videoconferencing systems connected to the digital switched public telephone network. Chapter 18 describes methods of combining participants for multipoint conferences.

■ *Computer-based systems using T.120.* The second category of computer-based desktop systems employs boards that conform to the international standards for visual telephony. The computer is connected to the telephone network using the same digital telephone lines as traditional videoconferencing products. Just as the LAN products allow users to work interactively on computer files, a new ITU standard known as T.120 enables computer-based systems that connect via the telephone network to share files and interactively work on the data in those files while videoconferencing. Because these products use advanced digital telephone services, they typically include call management features such as calling party identification, call history, and automatic redial.

The interactive sharing of files is the capability that distinguishes both types of computer-based videoconferencing products from traditional visual telephony products. People separated by miles can work together as though they were at the same desk. One such application is telecommuting. Other new applications that tap both the interactive video and data transfer capabilities of these products will probably bring a host of new applications to the field of visual telephony.

Computer-based desktop products cost between $2,000 and $8,000 per site.

Computer-based visual telephony was designed to be used by individuals. Later in this book we will discuss the relationship between video quality and the amount of telephone transmission capacity used to carry the video information. For now, let us simply say that the less transmission capacity available, the poorer the video. Most computer-based products were designed to use a relatively small amount of transmission capacity. The picture is still acceptable when the image is a head-and-shoulder shot of one person. This is particularly true when the picture is viewed in a small window.

Many organizations are not able to justify the expense of equipping a large number of desktop computers with videoconferencing boards and providing them with digital access lines. This is particularly true when potential usage for each individual is low. So instead, organizations might equip one computer and establish it as a shared resource. With several users sharing the system, the expenditure involved is easier for companies to justify.

Because the computer-based videoconferencing installation will be used by several people, it is possible that three or four people will participate in the same videoconference. In this case, the system can be fitted with a larger monitor and a camera with variable focal length. The system can also be connected to multiple telephone lines so that better picture quality is available. (See Chapters 8 and 9 for information on monitors and cameras; see Chapter 17 for a detailed discussion of transmission options.)

This configuration is a very cost-effective small group system. In fact, as computer-based systems improve in quality and decline in price, this type of installation may replace small group systems.

One caveat is in order. Computer-based systems were designed for use by one person. Files stored on the computer hard drive remain there after the system is powered down. Small group systems were designed as shared resources. Information stored on small group systems during a conference is deleted when the system is turned off. For security reasons, a shared computer-based system should be connected to a LAN. Users should be instructed to save files to their LAN directories, rather than to the system's hard drive.

3.3 Selection criteria

Each type of system—custom-built room, rollabout, desktop system, computer-based system—has its own benefits and drawbacks. The most obvious differences are price and size. But there is more to it than that. More significantly, each system differs in features and capabilities. When implementing a visual telephony system, the most important issue is how the system will be used.

3.3.1 What takes place during an in-person meeting?

People approach change with skepticism. The best way to introduce a new way of doing business is to minimize the amount of change people are asked to accept. It may be tempting to use the introduction of videoconferencing as an opportunity to modernize completely the way an organization functions. But the result of such a drastic change may be that the user community simply resists the use of videoconferencing.

Instead, to introduce videoconferencing successfully, study the way meetings are currently conducted. Ask yourself these questions: How many people attend? How many travel to attend? How many participate? How many are observers? What presentation aids are used?

- *The number of attendees.* As noted in Chapter 1, unlike travel, there is no cost for having extra participants in a videoconference. Observe how many people attend a meeting when it is held at their work locations. This is the number that is likely to attend a videoconference. Plan all videoconference rooms to accommodate the number of on-site people likely to attend.

- *Nonactive participants.* Not all attendees actively participate. In some meetings, people are invited to observe and learn about the subject matter of the meeting. It may be feasible to have seating for active participants and additional seating in an observation section.

- *Meeting activities.* Observe what takes place during a meeting. What presentation aids are used? It is tempting to disregard the occasional presentation using 35mm slides as an anachronism. However, the cost of providing the ability to display slides during a videoconference is small, particularly when compared to the benefits derived from using videoconferencing. To the user community, the benchmark against which a videoconference is measured is a conventional in-person meeting. If a user cannot use his or her typical meeting graphics in a videoconference, he or she is likely to reject it in favor of an in-person meeting.

- *Corporate culture.* One of the trickiest aspects of designing an organization's network of videoconferencing sites is dealing with the corporate culture. Earlier, we noted that a rollabout system provides the same capabilities as a custom-built room. Yet, a site-built conference facility is far more costly. In spite of this, many site-built rooms are installed every year because they are perceived to better suit the corporate culture.

In one company, senior management was concerned that employees would view videoconferencing as simply cost cutting rather than the intended objective of increasing efficiency and productivity. To dispel this idea, very well appointed site-built videoconference facilities were located in every regional office. Conversely, in another organization, videoconferencing was instituted as part of an overall "belt tightening." Expensive videoconference facilities would send the wrong message in this instance. Top management therefore required that all systems be rollabouts and that a minimum of decorating be done in the videoconference rooms.

Not all mandates are as sensible. Businesses that strive for a flat organizational structure may decide to install the same systems in all locations. This may mean spending more than necessary at some locations and "making do" at others. Hierarchical entities may strictly mandate the types of systems that can be installed—headquarters has the biggest and most ornate room, regions have large rollabout systems, and districts have small systems—regardless of the actual user requirements.

It is important to design videoconferencing facilities to meet the needs of the user community. If you must follow mandates such as those just mentioned, compromise carefully. If possible, avoid concessions that affect system functionality. For example, you could give up an elegant custom-built table in favor of a folding table that is just as functional. In contrast, giving up a large monitor in favor of a smaller one could negatively affect the user's ability to see distant participants. If possible, let the usage determine the features and functionality of the equipment provided.

3.3.2 Avoiding the pitfalls

The determination of the right type or types of videoconferencing system(s) is a critical part of implementing a successful network. Here are some of the pitfalls to avoid:

■ *Price.* Some systems may achieve low market prices by sacrificing quality. Users may not be able to conduct business with poor-quality video and audio, and the low-priced system may go unused. Unfortunately, higher prices do not guarantee better performance. System capabilities that are not used add nothing to a system's value. The selection process should therefore include a careful assessment of needs as well as a consideration of price. We discuss the various elements of a system in detail in later chapters to help make your selection of the right balance between price and performance much easier.

■ *Latest technology.* There will always be a better, cheaper product around the corner. That is the nature of an evolving technology. In laboratories, there are videoconferencing systems using holographic images that appear in a block of transparent plastic. In time, these lifelike images may be seated across the conference table from you. There are handheld computers that can receive and transmit video and audio over cellular connections. All of these innovations are exciting. However, they are not yet commercially available. The systems available today are reliable and cost effective. Almost any organization that frequently communicates with people who are geographically separated can benefit immediately from implementing current technology. Chapters 16 and 17 discuss international standards and the role they play in helping to smooth the path from current technology to future system offerings.

■ *Grander may not be better.* Some videoconferencing installations are in magnificent conference rooms with expensive furniture, complex audio/visual features, and expensive appointments. Such elegant rooms might inhibit potential users. On the other hand, a desktop system stuck in the corner of a busy office may never be used. There is a correct solution for each videoconferencing application. In fact, the same organization may need several different types of systems. The decision should be based on the requirements of the organization and the potential users.

3.4   Summary

Many different types of visual telephony systems can be used for videoconferencing (Table 3.1). Key points to be considered when selecting a system are the number of people that will be using it, whether or not computer applications such as file sharing are important, what additional audio/visual equipment might be used, the space available, and, finally, the budget. Information in Parts 3, 4, and 5 of this book will help you to select the right kind of system for your particular application.

Table 3.1
Summary of Videoconferencing System Types

System Type	Number of People	Strengths	Weaknesses	Price Range
Custom-built room	6 to 8 active participants, additional people in gallery	Can be designed for specific needs, with appropriate A/V support equipment	Equipment is not easily relocated	$100,000 and up; most around $250,000
Rollabout: large group system	6 to 8 active participants, additional people in gallery	Faster and less costly installation and more easily relocated than custom rooms	Some configurations not possible	$25,000 to $75,000
Rollabout: small group system	Up to 4 active participants	Cost effective for smaller groups	Fewer features than large group systems	$10,000 to $30,000
Desktop system	1 or 2	Takes up less space than rollabout units	Few options for additional equipment	$2,000 to $8,000
Computer-based system	1	Ideal for collaborative computer work	Not a shared resource	$2,000 to $8,000

4

Distance Education

THE TYPICAL DISTANCE education installation is an electronically enhanced classroom. These classrooms are constructed to serve a range of applications. Many distance education systems are installed to connect a college's or university's multiple campuses. Some high schools in large school districts share teachers, especially for advanced courses, by connecting high school classrooms at various schools throughout the district. Because the classrooms may be origination sites for some classes and receive sites for others, the visual telephony installations in each location are identical.

Other distance education installations are used to deliver instruction to off-campus sites that never originate instruction. For example, some state universities use distance education to teach students who cannot attend school on the main campus. The students attend classes at smaller local campuses near their homes. Were it not for distance education, they might not be able to take these courses. One North American university offers master's degree programs to students in the Republic of China. The entire degree program is delivered via interactive distance education. Many community colleges offer advanced classes to local high schools or offer specialized training and continuing education to people at their workplaces. In each of these instances, the instructor's originating system is different from the system installed to receive instruction in the remote classroom.

4.1   Configurations

The examples in the preceding paragraphs represent three different kinds of systems used in distance education: *origination only, receive only,* and *combination systems.* We discuss next the basic requirements for each of these distance education systems. Chapter 12 goes into more detail, describing various kinds of additional equipment that can be useful in distance education facilities.

4.1.1   Origination only

The university that is offering a master's program for Chinese students via distance education installed an electronic classroom that is dedicated to this use. Instructors use the room to teach students in Beijing. The room is never used to receive course material.

For this application and others that originate instruction but do not receive instruction, there must be a camera at the instructor's location to capture the instructor's image and a monitor to display the students in the remote classroom(s). The instructor must have a microphone to transmit her voice and a speaker with which to hear remote students. Of course, there needs to be a two-way transmission channel to connect the sites. These are the basic requirements, but you should realize that there are probably no installations this modest.

Distance education sites usually contain ancillary equipment (Figure 4.1). For example, the traditional chalkboard and chalk are usually replaced by a tablet and stylus. The instructor writes on the tablet with the stylus, and students at all locations see the annotations displayed on a monitor. Prepared instructional material is often imaged on a document stand, which is a camera attached to an imaging surface. A VCR and computer may also be used to send visual information. A

Figure 4.1 A rollabout instructional unit. This unit accommodates multimedia teaching aids and is easily moved from classroom to classroom. (*Source:* PictureTel Corporation, Andover, Massachusetts.)

facsimile machine might be used for the exchange of written assignments. Finally, there is a controller for call establishment and system management.

A system designed for originating instruction may or may not include provisions for local students. In some instances, the instructor teaches a local group of students and simultaneously provides instruction to students at remote classrooms (Figure 4.2). In a sophisticated design for both local and remote students, students at any location can be seen

Figure 4.2   A distance education facility. This site incorporates a rollabout teaching podium with a touch-sensitive control system for managing a variety of multimedia inputs. (*Source:* NEC Video Works, Irving, Texas.)

and heard by the other students when interacting with the instructor. In this case, the origination room would provide additional equipment so that not only could the instructor see his or her remote students but local students could interact with remote students as well.

4.1.2 Receive only

If a location will never initiate instruction, it can be more economically designed as a receive-only site. For example, the students in China are seen and heard by the instructor in the United States, but no instruction ever originates from the Beijing site. Their system is inter-active but does not contain the additional equipment required to support origination. At a bare minimum, the receive site needs a monitor, camera, microphone, and speaker.

Although a single monitor can be used to display prepared instructional material and to show the instructor, typically two monitors are installed. With two monitors, students can observe the instructional material while simultaneously watching the instructor. Chapter 9 discusses display system configurations in more detail.

A camera images the students so the instructor can see them. There should be one microphone for every three or four student seats so the instructor can hear the students. Finally, the students require a speaker so they can hear the instructor.

A VCR can be added to record the class for future review, and a facsimile machine might be located in the classroom so students can send assignments to the instructor.

4.1.3 Originate/receive

The high school classroom mentioned earlier is linked to a county-wide distance education system. It is used both as a receive site for courses taught by instructors located at other schools and as an origination site for courses taught by local teachers. It must function both as an origination site and as a receive site without reconfiguring equipment (Figure 4.3). As such, it must contain all the components described earlier for rooms that originate and rooms that receive instruction.

Practically speaking, a classroom that can be used for both originating and receiving instruction differs from the origination classroom only in that it must have a monitor for the instructor's image when the site is a receive location. If a monitor is located behind the instructor's podium, the classroom will be suitable for both a receive and origination classroom.

4.2 Distance education installations

In the previous chapter, we discussed the various kinds of systems available for videoconferencing applications. A variety of systems are also available for distance education.

4.2.1 Custom-designed classrooms

The vast majority of distance education classrooms currently in use were custom designed and built (Figure 4.4). Just as in videoconferencing, in

Figure 4.3   An originate/receive distance education classroom. This teaching podium allows instructor control of the teaching/learning experiences and incorporates a variety of multimedia peripherals. (*Source:* Tandberg, Herndon, Virginia.)

the initial stages of distance education there were not enough systems deployed to recognize the essential components of a standardized system. This is changing. With the large number of systems now deployed, a few system integrators have begun to offer standardized designs that can be custom-built into classroom space. These designs are more elaborate than most videoconferencing installations. Prices range from $40,000 to $150,000.

4.2.2   Rollabout systems

A basic receive-only classroom for small classes of students can be designed using a large group videoconferencing rollabout system. Applications such as continuing education furnished at the workplace and courses provided to prison inmates can be inexpensively equipped using an off-the-shelf system. Because these systems are transportable, the education provider can own the equipment and have it installed when

Figure 4.4 A custom-built distance education facility. This lecture hall was custom designed and built for a university distance education program. (*Source:* VTEL Professional Services, King of Prussia, Pennsylvania.)

courses are contracted. This can give the education provider a competitive advantage.

The typical videoconferencing rollout can be modified for larger classes of students at receive-only sites by adding a microphone mixer and several microphones. This is a simple modification, and more and more units are likely to be sold for this purpose.

Rollabouts can also be modified for origination-only sites or for sites that both originate and receive. One such modification might be the addition of an infrared tracking system that keeps a pacing instructor in view. The cost of modified rollout systems can range from $15,000 to $90,000.

4.2.3 Desktop systems and computer-based systems
The earliest distance education applications were initiated by technology companies, and today many corporations provide continuing education

and training to their knowledge workers with dedicated communications links between the company's facility and a university. Professionals attend classes at their work location, usually by going to a specially equipped classroom. Desktop videoconferencing, and particularly computer-based conferencing, offers the *potential* for delivering classes to students right in their offices or cubicles. However, distance education via the desktop is not widespread.

The technology has long existed to allow educational material to be broadcast to people at their work location. Many organizations already do this, and some companies have enabled their knowledge workers to access stored computer-based or video instruction from their workstations. The main stumbling block for interactive distance education at the desktop is the inability to link a large number of desktop systems while still giving the instructor the ability to interact with the students. Multipoint control units allow people to see and be seen by others on the same multipoint conference. (See the discussion of multipoint visual telephony in Chapter 18.) However, if many individuals are connected, it becomes difficult, if not impossible, for the instructor to manage the interaction. And, practically speaking, interaction with large numbers of students must be limited. The alternative, teaching in more of a lecture style to a large number of students who are not in view at all times, is not always comfortable for instructors or desirable for the learning experience.

Another problem with distance education to the workplace via desktop or computer-based systems is that adult students at work may not have ideal environments for learning. Coworkers and telephone calls may interrupt the student's concentration. Ringing telephones and background noise may be transmitted along with the student's audio. There may not be sufficient space on the desktop for another piece of equipment.

Distance education installations that bring instruction to student desktops in the schools are not yet a reality. The cost of equipping each desk with a computer is high, and it is difficult for an instructor to interact personally with each student.

However, there are computer systems that support large-group distance education systems. These may involve individual computers or simpler response devices at each student's desk. In these systems, the origination site can pose questions to the distant students, who can respond via their desktop systems. The student units are connected via telecommunications to the instructional site. The responses are collected by the instructor's computer, which records them and presents a summary chart for the instructor. If only a few students respond incorrectly, the instructor can engage them in a dialog. If there is a higher level of incorrect responses, the instructor is prompted to review the material. This is a simple example; more sophisticated programs for computer-assisted education are in use at distance education installations.

Development of multipoint control units that enable users to exchange computer data during a session, the falling costs of computer hardware, and the general need to reach students in widespread locations will all spur the development of computer-based distance education. Clearly the use of computers is a trend in education and it will continue to have a major effect on distance education as well.

5

Telemedicine

INTERACTIVE VISUAL TELEPHONY has been used by the medical community for many years. Medical schools, health maintenance organizations, and pharmaceutical companies were among the first organizations to embrace visual telephony.

Medical schools use distance education classrooms to teach a wide variety of courses, from typical subjects such as nursing, biology, chemistry, anatomy, and physiology to more advanced and specialized medical courses. Even though visual telephony is used to deliver these courses to students studying health care, the application is better characterized as distance education than telemedicine. We discuss distance education and typical installations in detail in Chapter 4.

Health maintenance organizations (HMOs) began using visual telephony to improve patient care and contain cost outlays long before there was national awareness of spiraling health care costs. The HMOs used visual telephony for training health care workers on everything from the use of new equipment to filling out government reports. Once again, the use is more accurately categorized as distance education.

International pharmaceutical companies were also among the first corporations to adopt visual telephony. They use it often to present new products to groups of doctors at hospitals or medical centers. Though the presentation may include ancillary equipment unique to the medical community, the application is, in fact, videoconferencing, which is discussed in Chapter 3.

Until recently, visual telephony was not widely used for patient care. The true power of this technology is only now beginning to be applied to the delivery of health care. Today, telecommunications is common in three types of applications: teleradiology, remote diagnostics, and remote case management. Visual telephony, specifically, is used in two of these, remote diagnostics and remote case management.

5.1 Teleradiology

Teleradiology itself does not use visual telephony channels, but it may be included as part of a remote diagnostics system that does make use of visual telephony, or teleradiology might be enhanced with visual telephony to connect specialists during a consultation. For this reason, it is important to understand more about teleradiology and its connection to remote diagnostics and videoconferencing.

Teleradiology involves the transmission of images created from X-rays, magnetic resonance imaging (MRI), computerized axial tomography (CAT) scan, and ultrasound equipment. In most cases, teleradiology does not rely on visual telephony for the transfer of information. The American College of Radiologists has produced standards for the transmission of X-ray images over telephone lines. The images are very detailed and require that a large amount of information be transmitted in order to replicate the image on the distant end. The radiologists' standard does not allow data compression for fear that the process might eliminate meaningful visual information. It also requires that images be displayed on special monitors that provide more resolution than a standard video display. X-ray images, used for diagnoses, are therefore transmitted independent of the visual telephony connection.

Other images, such as ultrasound, MRI, and CAT displays, can be sent as visual telephony images. In particular, ultrasound images are normally displayed on monitors similar to ones used for videoconferencing.

There are a number of teleradiology installations that do not incorporate visual telephony. Both the radiology department and the radiologists on call have a specially equipped IBM-compatible or Macintosh computer connected to telephone lines via a modem. The radiology computer has a scanner that digitizes an image from the X-ray film. Software stores the image and transmits it to the physician. The software has error detection and correction. The doctor reads the X-ray and calls the radiology department with his or her assessment. If normal telephone lines are used, the file takes 10 minutes or more to be transmitted. ISDN channels can cut the transmission time in half.

Computer-based visual telephony boards and software can be added to the computer to allow the parties to hold face-to-face discussions when the file transfer has taken place. However, more often, a teleradiology computer and software are integrated into a telemedicine system. X-ray, MRI, CAT, and ultrasound images are sent along with live video information, over digital telephone lines.

The use of teleradiology is limited in part by the special equipment needed, but more by the reality of the practice of medicine. The fact is that any medical facility that has the ability to take an X-ray will have access to radiologists who can read them. Rarely does a radiologist need assistance from a specialist to interpret a film. In these circumstances, either the X-ray can be delivered by courier to the specialist or the patient can be transported to a medical center that has experts who can diagnose the problem. The same can be said for the output of MRI equipment and CAT scans, since this expensive equipment is usually installed in or near a hospital. There is little or no need for distant physicians to collaborate on a diagnosis.

In its limited uses, teleradiology seems to have two contrasting applications. The first application is usage within a *local area*. In many parts of the country, doctors live an hour or more from the hospital at which they practice. Patients who come to the trauma care unit on weekends and during the night are X-rayed by a technician, and an on-call physician is contacted to come read the X-ray. The doctor may drive a long time simply to confirm that the patient did not break a bone. Teleradiology allows the physician either to stay at home or travel to his or her office to receive the images, reducing the actual time required to make this diagnosis.

The second application is the provision of specialized care to *remote parts of the world* that do not have access to expert medical care. The most obvious example of this is military field medical units. Remote areas of developing nations provide another example.

5.2   Remote diagnostics

More important today is the remote diagnostics system, which frequently includes a computer running teleradiology software.

5.2.1   Expert opinion

There are many circumstances in which appropriate medical resources are not readily available to the patient. People living in rural settings may be far from physicians with specialized knowledge. Military personnel in primitive postings may not have ready access to doctors with the required skills. The cost of taking a dangerous prisoner to a hospital can often exceed the cost of the medical treatment. These are among the situations that can be addressed by a remote diagnostics system.

Remote diagnostics systems are used after the patient is first examined in person by a health care professional. In some cases, the patient is seen by a physician. In other situations, a nurse practitioner or physician's assistant is the first to see the patient. In the military, medics provide the initial patient assessment. The remote diagnostics system is used when there is an indication that more specialized knowledge may yield a better diagnosis and more effective treatment. The system is able to link the patient and local health care provider at one site with a consulting physician or team of physicians at another location.

5.2.2   Equipment

Few systems are designed for use in operating rooms or other places in which volatile gases are present. The typical installation is in an examination room. Transportable carts are strongly favored for this application (Figure 5.1). They can easily be shipped and installed in a

Figure 5.1 A telemedicine cart. This portable unit can be configured with a variety of medical devices to suit the application. (*Source:* NEC Video Works, Irving, Texas.)

medical facility. In contrast, building a custom room is time consuming, costly, and lacking in flexibility.

The systems used for remote diagnostics need not be the same at both the consulting physician's site and the patient location. However, in the majority of installations in use today, the two systems are alike. There are two reasons for this. First, in the case of HMOs, the same site may be used by a consulting physician and later as a patient location. Hospitals within the network have areas of expertise that are shared among member institutions. Secondly, it is easier for the doctor to work with a remote system that is the same as the one he or she is using locally.

Remote diagnostics applications can make use of rollabout systems or components configured on transportable carts that are designed to be easily set up in existing examination rooms. Often many of these systems are deployed throughout a telemedicine network.

The typical remote diagnostics system includes a large monitor on which the examining physician can view the patient, X-rays, and other diagnostic material. The examining physician is imaged with a camera located near the monitor (Figure 5.2).

At the other end, the patient and local doctor, nurse, or other health care professional can see the examining physician on a monitor. The image of the patient that they transmit is usually viewed in a small window on the same monitor, a feature called *picture-in-picture* (discussed in more detail in Chapter 9). The images on the local monitor and window can be interchanged. In this way, one monitor can show both incoming and outgoing views.

In the patient's location, there is a camera on a stand, similar to the document camera used in videoconferencing, that can transmit

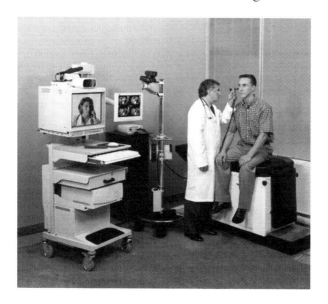

Figure 5.2 A visual telephony system for tele-health care. Systems such as this can be used for remote diagnostics, consultation, medical education, and other applications. (*Source:* Tandberg, Herndon, Virginia.)

X-rays and diagnostic material. As noted earlier, X-rays cannot be read by a radiologist using compressed video; however, once a diagnosis is made, an X-ray shown on a video system can be used to convey information graphically.

There is also a tripod-mounted camera that is used for close-up images of the patient's body. Both locations can select which camera output is transmitted, and both can control the tripod camera to give the best view of the affected area.

There is a microphone near the monitor and one or two others installed in the room. There is also a speaker. For confidential discussions, the speaker can be disabled and the doctor can use a headset instead.

A teleradiology computer and special data transmission lines may be incorporated into a remote diagnostics system, allowing a radiologist to make remote diagnoses.

5.3   Remote case management

Following a stay in the hospital, patients often require follow-up care. Family members can assist. Health care workers can come to the home to provide care that is beyond the ability of family members. In many instances, a visiting nurse must make daily visits to assess professionally the patient's condition.

5.3.1   Medical monitoring

Desktop visual telephony systems are ideal for remote case management of patients at home or in distant clinics. A system can be placed near the patient's bed. Devices for monitoring the patient's blood pressure, pulse, temperature, and heart can be connected to the data ports provided on these units.

Routine checks can be initiated by the monitoring center to check vital signs and the patient's general well-being. At prearranged times, a nurse will call the patient and ask the patient to assist by putting an arm in the blood pressure cuff, a finger in the pulse cup, and a temperature probe under the arm or in the mouth. The nurse can also ask the patient to place the electronic stethoscope on his chest. The video connection can be used to assess skin color and to time respirations. The video not only carries images of the patient to the nurse but allows the patient to see and talk with the nurse. The patient can hold a private conversation with the nurse or other medical personnel.

If they can be monitored at home, patients can be discharged from the hospital sooner. Not only does this improve the morale of the patient and his or her family, but it improves outcomes by reducing patient exposure to infection. The use of visual telephony can also reduce costs by decreasing the number of visiting nurse calls that would be made while the patient recuperates.

Over time, a rapport can be established between the patient and nurse. In a hospital setting, it is often necessary to vary patient assignments daily. Patients are assigned to rooms based on the availability of beds. Nurses are assigned based on case load and the proximity of patients. The electronic assignment of cases is not affected by the location of the patient. A nurse can be assigned to a case and follow the patient's progress throughout the patient's recovery. The same nurse might be the person who makes weekly in-person visits.

In emergencies, the patient can push a call button to connect automatically to a monitoring center for assistance or for expert advice. When a patient initiates a call, it is routed to an emergency response site. Doctors or nurses are immediately available to accept the call and respond. The equipment is identical to the nursing site system. Because the caller information is available on an ISDN call, it can be used to bring the patient file to the data monitor. The emergency staff can either dispatch an emergency response unit or transfer the call to the nursing site.

If an emergency response unit is sent to the patient site, the emergency personnel can stay on the line and try to calm the patient. The audio can be relayed to the response unit so that the personnel will be aware of the situation when they arrive. The link can be maintained until the patient is transported to the hospital. In this way, the emergency room staff at the hospital is better prepared to handle the situation when the patient arrives.

The ease with which knowledgeable medical personnel can be contacted, night and day, can help ease the anxiety of older patients and people with potentially life-threatening conditions.

5.3.2 Typical installations

The desktop videoconferencing systems described in Chapter 3 are easily adapted to remote patient monitoring.

Nursing site

At the nursing site, there is typically a 20-inch monitor to display the patient's image. A second data monitor initially displays key pieces of patient information. The nurse may also review the entire patient record using this data terminal. The nurse initiates the call by clicking on an icon.

During the electronic visit, the information from the remote sensors (e.g., temperature, pulse, blood pressure) is displayed on the data terminal. The nurse can return to the record at any time. By accessing a report page, the vital signs recorded by the remote sensors are entered in appropriate blocks. The nurse can also make comments that will be added to the permanent record and added to the initial display screen in the future.

A small camera located adjacent to the video monitor images the nurse. A microphone on a flexible holder is adjustable, so that the nurse can speak to the patient in a conversational tone of voice and still be heard clearly. The nurse can listen through the monitor speakers or with an ear piece that ensures patient confidentiality.

The nurse also has a standard telephone nearby to summon emergency medical personnel to the patient's home if needed.

Emergency response site

The emergency response site is equipped the same as the nurse's site, allowing emergency medical personnel to call up records and to access vital signs in real time.

Patient site

At the patient's location, the videoconferencing unit is placed on a table near the patient or on a specially designed cart that incorporates the videoconferencing unit and remote monitoring equipment for blood pressure, pulse, temperature, heartbeat, and other vital signs.

5.4   Practical uses of telemedicine

The popular press often features articles about future applications of telemedicine, including remote surgery. It is very hard to imagine the justification for transporting a sophisticated surgical system to a remote site so that surgery can be performed on someone, rather than transporting the patient to the physician. Perhaps when people inhabit space stations deep in space, this will come to pass. Until then, there are many more practical uses of the existing technology for bringing quality health care to the people of the world.

Today, physicians with particular knowledge use telemedicine to assist local doctors in performing surgical procedures. People are treated for emotional disorders over a telemedicine system. Patient progress is monitored over digital telephone lines. Traditional distance education is used to provide hospital staff with continuing medical education. Administrators use videoconferencing to efficiently provide health care to the communities they serve. In all of these examples, the distance is bridged using telemedicine, providing more opportunity for quality health care.

6

Law and Criminal Justice

VISUAL TELEPHONY IS used in law and criminal justice for meetings and for training, which are videoconferencing and distance education applications, respectively. There are applications unique to this environment, too.

6.1 Typical applications

Some of the more common uses of visual telephony in law and criminal justice include video arraignments, depositions, testimony, and parole hearings.

6.1.1 Video arraignments

A person arrested without a warrant is entitled to appear before a judge and have the charges against him or her read in public. At that time, the accused pleads either guilty or innocent. This is known as an *arraignment*. The entire process takes less than 15 minutes. The most common criminal justice application of visual telephony is video arraignments. It costs thousands of dollars to transport a prisoner from a detention center to the courtroom. A video arraignment reduces the cost to a few dollars.

During an in-person arraignment, the defendant appears before the judge. The prosecutor and defense attorney are also present. There is no jury or witnesses. A very basic videoconferencing system is all that is required to replace this proceeding with a video arraignment.

6.1.2   Depositions

In preparing for trial, whether criminal or civil, attorneys frequently question potential witnesses under oath. The witness statement is a *deposition*. Both the plaintiff's attorney and the defendant's attorney are usually present while a deposition is being taken. If the witness is not a local resident, both attorneys may have to travel to take the deposition. Alternatively, one of the principals in the case may pay the witness's travel expenses. In either case, this expense can be avoided by using visual telephony. Taking depositions using visual telephony is the second most common application in law and criminal justice.

6.1.3   Victim and witness reports

It has been said that in America's large cities, a victim of a crime is victimized twice: once by the felon and another time by the system. The victim is often asked to return to the police station near the crime scene to provide details to the police investigating the incident. When a suspect is arrested, the victim must travel to the police station where the suspect is being held. The victim must also meet with the prosecutor. Finally, the victim must testify. These many trips can take hours out of a busy person's life. People have been known to drop the charges rather than continue with this time-consuming process. By linking police stations using visual telephony, the victim is spared countless hours of travel. Instead, he or she travels to the nearest station house and uses the visual telephony link.

6.1.4 Testimony

Under some circumstances a person can testify using visual telephony. Several witnesses have given testimony from their hospital beds. Children have been allowed to testify over video, rather than being exposed to the frightening environment of a courtroom. Circumstances may also be such that a prisoner is allowed to give testimony from the jail. In one widely publicized situation, a person was subpoenaed to testify in a civil case. However, the person was under indictment for criminal activities related to the case and had fled the country. He would have been arrested had he appeared to testify. Instead, he agreed to provide testimony over a video link.

6.1.5 Parole hearings

The use of visual telephony does not end when a person has been sentenced to prison. When the prisoner is eligible for parole, the parole hearing can be conducted over a video link. The travel costs of the parole board are avoided as well as the expense of heightened security while the parole board members are at the prison.

6.2 Typical installations

Most judicial applications are meeting applications with some added requirements. As such, they are generally well served by typical videoconferencing systems (see Chapter 3 for details about videoconferencing), with some modifications.

6.2.1 Desktop systems

Desktop systems are ideal for connecting the victim with investigating officers and the prosecutor. They can also be used for witness identification of a suspect.

6.2.2 Small and large group rollabout systems

Depositions can be taken using either a small group system or a large group system, depending on the number of participants. Refer to the discussion in Chapter 3 about the number of participants suitable for small and large group videoconferencing systems.

One addition that would be made to a typical small group or large group videoconferencing system used for depositions is the inclusion of a VCR that continually records the date and time, along with the audio and video. Although this is not always required, in some cases, VCR tapes have been admitted into evidence. The date and time stamps are required by some courts.

Parole hearings can be held using a large group rollabout at the parole board location and a small group system in the prison.

6.2.3 Enhancements to courtroom systems

Video-enhanced courtrooms can be used for trials, hearings, depositions, arraignments, and other applications. Many court systems are designed to support all potential judicial applications. At any given time, a feature not needed for an application is simply turned off. For example, although a system has a monitor that allows the jury to view a person testifying over video, it can be turned off when the system is being used for video arraignments and there is neither a jury nor anyone testifying.

Designing a visual telephony system for a courtroom is challenging. Either large group systems are modified for the applications, or equipment is custom built into the courtroom. Let's look at how a system might be designed for a defendant who is tried from prison.

The defendant in a civil or criminal trial who is tried from prison rather than in the courtroom has the right to look at the judge, the prosecutor(s), the defense attorney(s), and the witness(es). Cameras must therefore be placed accordingly to image the judge, the attorneys, and witness stand. The camera focused on the judge can be mounted on the wall opposite the judge or attached to the judge's monitor. Likewise, the camera for the attorneys can be wall mounted or affixed to the monitor they view. The camera for the witness stand is usually wall mounted.

The system is designed with a video mixer so that the defendant can view all four images. The mixer has a feature that takes four video inputs and places them in quadrants to be transmitted as one picture. The prisoner sees the four images simultaneously.

Items offered into evidence during a trial may include documents, photographs, objects, presentation boards, and easel pages. A document camera with a camera mount that can be tilted up to focus on an easel is usually provided in a judicial visual telephony system. When presentation material is displayed, the quadrant feature is turned off. The material is displayed on a full screen.

The judge, the prosecutor(s), the defense attorney(s), the jury, and the court clerk all have requirements to see the accused person. The judge is usually provided with a personal monitor about 9 to 12 inches in diameter positioned either on the judge's bench or built into it. The judge is usually provided with a selector so that he or she can choose to view the prisoner on a full screen, the four quadrants, or the output of the document camera (this provides preview capability for material that is to be presented). A ceiling-mounted monitor is provided for the jury. It will always be used to show the prisoner. The prosecutor(s) and the defense attorney(s) can both watch the defendant on a monitor placed in front of the judge or mounted on the wall behind the judge.

Typically a clerk operates the courtroom visual telephony system. The clerk needs a preview monitor to be able to switch the system correctly and focus the document camera. This is usually a monitor 15 inches in diameter.

Audio is the most difficult part of court systems. The judge and attorneys must be able to speak normally and be heard by the distant participant(s). The courtroom participants (judge, prosecutor, defending attorney) are provided with individual microphones. The judge's microphone is usually a tabletop model, mounted on the monitor or on the bench. Tabletop microphones, placed on their respective worktables, are provided for the two attorneys. Everyone must be able to mute the microphone so that a private conversation can be held near it. Each microphone should have a light that indicates when the microphone is active.

The defense is also provided with a means of speaking privately with the prisoner. One popular method is a telephone located at the defense table. When the attorney lifts the receiver, the defense attorney's audio is sent only to the detention center. The prisoner hears both the courtroom audio and the attorney. The defense attorney can make private comments to the prisoner, without being heard by anyone else or taped.

Courtrooms are often acoustically challenging. The audio system must compensate for the acoustic deficiencies of the room and also control echo or feedback.

Videotapes are usually made of legal proceedings. Both the courtroom and prison audio are mixed on the tape. Only one site's video can be recorded. The courtroom video is generally the source video; however, the clerk must switch to the video from the prison when the defendant is testifying.

6.2.4 Equipment modifications for prisons and detention centers

Equipment used in detention centers is very simple, usually consisting of a small group rollabout system or equipment housed on a cart. The monitor has a built-in speaker. A camera on a fixed mount, with manual zoom and focus, is located just above the monitor. Enhanced prison systems may include a camera mounted on a pan/tilt head and a lens that can be remotely zoomed and focused. The camera will likely have an autofocus lens and be controllable from an infrared handset in the prison or remotely from the courtroom.

A special requirement for systems placed in detention centers or prisons is that they be encased in a sturdy, washable enclosure with a Plexiglas front. The system enclosure should be able to withstand being shoved and pushed violently. Ceiling-mounted microphones are the easiest to maintain. There should be remote diagnostics through a dial-up modem. All cabling and electrical connections should be routed through sturdy conduit.

7

Sales and Marketing

GIVEN THAT SALES and marketing are intense areas for interpersonal communications, it is no surprise that visual telephony is heavily used in these environments.

7.1 Applications and installations

Meetings to plan marketing strategy or to review advertising campaigns, or training sessions on sales skills or new products are best described as videoconferencing or distance education applications. In addition to these popular applications, there are others unique to the sales and marketing field.

7.1.1 Market research

One nice feature of visual telephony is that it is interactive. This makes visual telephony ideal for various market research applications, including focus groups.

Typical focus group facilities are built with a one-way mirror separating a conference room from an observation area. The firm contracting for the research may have a representative attend the focus group sessions and observe from behind the mirror. After each session, the representative can discuss the findings and ask the moderator to pursue additional information during the next session.

Because it is not always possible to have a client representative present during the sessions, many focus group meetings are videotaped for later review with the client. An alternative is to place a small group video system in the conference room so that the client can observe the sessions without traveling to the focus group facility.

Another marketing research application involves interaction with shoppers in a shopping mall or supermarket. A small rollabout can be installed so that the client can actively discuss products under consideration with potential buyers. The same system can be used to pretest commercials. The client can watch people's reactions as they view a commercial displayed on the system's monitor.

7.1.2 The virtual branch office

Kiosks located in office buildings, shopping centers, and within large stores can be equipped with video and used for consultative selling (Figure 7.1).

Consultation with potential customers, reservations, and ordering are three of many applications. Travel agents, hotels, rental car companies, and many other service organizations can extend their markets very inexpensively with visual telephony. Banks can bring full services to all branches, even little ATM locations in supermarkets and shopping malls.

A video-equipped kiosk for banking applications would have an interactive system that allows a potential customer to find out the current rates for car loans, mortgages, certificates of deposit (CDs), and so on (Figure 7.2). The video link can be called up by the customer on demand. For example, an inquiry about mortgage rates would present the customer with an option for discussing specific requirements with a loan officer. The video is shown in a window on the kiosk's monitor.

These kiosks are custom designed to meet the application requirements.

Figure 7.1 A visual telephony kiosk. Users can interact with financial services personnel using the visual telephony features in this kiosk. (*Source:* PictureTel Corporation, Andover, Massachusetts.)

Figure 7.2 Customer interaction via visual telephony. A customer in a kiosk and a banking adviser in her office discuss financial products. (*Source:* PictureTel Corporation, Andover, Massachusetts.)

Part 3

Equipment and Environment

EACH APPLICATION AREA—videoconferencing, distance education, telemedicine, law and criminal justice, sales and marketing—has unique design requirements for visual telephony. In the last several chapters, we introduced some generic designs and system features.

Now let's look at equipment options in more detail. Part 3 contains specific information about cameras in Chapter 8, display systems in Chapter 9, audio systems in Chapter 10, control systems in Chapter 11, and various other useful equipment in Chapter 12.

To be certain the various component parts of your system work as intended, you must give careful consideration to the environment in which they are installed. Chapter 13 includes useful guidelines for selecting and preparing your physical facility.

8

Cameras

VISUAL TELEPHONY ATTEMPTS to replicate face-to-face encounters (meetings, consultation, instruction, diagnosis, identification). The extent to which the equipment can re-create the in-person situation and the degree of comfort it provides to the user during a visual telephony experience will determine the success of the application. If people can't see or hear well, or if they can't manage the system and the task at hand easily, they won't use it.

The most obvious components of a video telephony system are the camera and display. Choosing the right equipment depends in large part on the number of participants using the system. The larger the number of active participants, the more difficult it is to design a system to meet the needs of the user community. The two most difficult problems are (1) imaging the active participants in a natural way and (2) ensuring that *all* participants, both active and observing, have an unobstructed view of the incoming video and visuals used during the session.

In this chapter we discuss options for the main video camera(s) and for additional cameras that are needed in many applications.

8.1   Camera features

Cameras are available with a range of imaging quality and capabilities. Unquestionably, a better camera provides a better image at the distant location. However, better cameras cost more. The overall quality of video images depends on the quality of the camera, monitor, codec, and the type and amount of transmission capacity used to convey the signal. The camera needs to be selected with the quality of these other components in mind. A very expensive camera might produce no perceptible improvement in a system using lesser components. For information on other components, note that Chapter 9 provides information on monitors, Chapters 14 through 18 provide a tutorial on transmission, and Chapter 16 is devoted to video codecs.

System designers strive to assemble the optimum components to achieve an established price. Their selections may not be right for every application. In some instances, such as computer-based videoconferencing, upgrading the camera and monitor adds greatly to the system's usefulness.

Price strongly indicates image quality, but it indicates other things as well. Spending more is not always advisable. For example, the difference between the image received from a broadcast-quality camera and one costing half as much does not justify the price difference. Broadcast cameras cost more for many reasons. One reason is that they have features valuable in multicamera television productions. These features have no value in visual telephony.

8.1.1   Autofocus and auto iris

For cameras used in group situations, lenses should have two automatic capabilities: autofocus and auto iris. During the session the camera might be moved or the participants might move. With autofocus, there is no need to refocus the camera.

The lighting in a videoconference room or classroom or other visual telephony venue should be even with no shadows. In reality, however, the lighting may be less than ideal, with some areas not correctly lighted. Furthermore, different clothing colors and textures absorb light differently. Auto iris keeps the images a uniform brightness.

Although autofocus and auto iris are essential lens features for larger systems, both the cost and the size of these lenses make them impractical for desktop and computer-based systems. In these designs, the camera is generally fixed in one position and does not have to be adjusted frequently.

8.1.2 Automatic white balancing

The camera should perform automatic white balancing. When a camera is turned on, it must be color corrected for the environment. An auto-white-balance camera can quickly self-adjust. This feature is necessary for cameras used in all visual telephony systems.

8.1.3 Single-chip versus three-chip cameras

Cameras today use computer chips called *charged coupled devices* (CCDs). A CCD converts an image from light into an electrical current that produces a video signal in the camera. Sophisticated studio cameras have three CCD chips, one for each of the primary colors used in video (red, green, and blue). Three-chip cameras produce superior images compared to single-chip cameras, but because they are intended for studio applications, they can seldom be fitted with an autofocus lens. They are also much more costly than cameras with a single CCD chip. Therefore, most visual telephony systems in operation today have single-chip CCD cameras.

8.1.4 Chip sizes

The size of the chip also influences the quality of the image. A 1/2-inch chip outperforms the 1/3-inch chip found in consumer camcorders. Because it is difficult to locate a 1/2-inch chip camera that has the desired visual telephony capabilities mentioned earlier, most systems have 1/3-inch chips. This may change as camera prices continue to decline and as competition for the growing visual telephony market causes manufacturers to improve their offerings.

8.1.5 Lenses

The lens used for group visual telephony should have a 60-degree pick-up field. Such a lens is often referred to as a *wide-angle lens.*

8.1.6 Resolution

The resolution from cameras used in group systems should be at least 525 vertical lines on 480 horizontal lines. Desktop systems will use cameras with resolutions in the range of 300 vertical lines on 350 horizontal lines. Video resolution is covered in detail in Chapters 9 and 16.

8.2 Placement of participant cameras

With correct camera placement, a person using visual telephony can look at someone in a remote site straight in the eyes, just as though that person were in the same place. Television news anchors, for instance, are trained to look into a camera to give you the feeling that they are talking directly to you. But focusing on a camera would not be natural for most of us. To create the illusion of eye-to-eye contact, the camera must be placed close to what the user would naturally be observing. In most instances, the user is watching a video display of the activities in the distant location. Therefore, in most systems, cameras that image participants are located near the video display. For a videoconference, the camera is usually mounted in a front wall or rollabout unit near the image of the incoming video.

The same placement is used for remote diagnosis. The camera that images the consulting physician is located near the system's main monitor. As the physician looks at the monitor to see the patient, the patient feels as though he or she has eye contact with the physician.

In a distance education classroom, the camera is often mounted on the back wall, behind the students who are present in the instructor's location. When the instructor is looking at the local class, he or she is also looking into the camera. In this way, the instructor appears to be speaking directly to all the students, both local and remote.

In video-enhanced courtrooms, four cameras are used, each mounted on the wall opposite the locations they are imaging. Typically, the judge's bench is centered along the front wall. Thus the camera that captures the judge's picture is mounted on the rear wall. Alternatively, it could be mounted on the judge's monitor. The camera that is used for the witness, who sits to the judge's left, is on the rear wall. The cameras for the prosecutor(s) and defense attorney(s) are mounted on the wall behind the judge.

In computer-based collaborative work, the camera often rests on top of the computer monitor.

8.3   Imaging one or two participants

In addition to the position of the camera, the number of cameras and the way the cameras are controlled are important design considerations. These issues relate directly to the number of participants.

If the application involves only one or two people seated together at a site, a single camera is sufficient.

Video telephones and other desktop systems generally have a single built-in camera. Usually the iris (brightness adjustment) and focus are manually controlled. The camera is set to image users at a predetermined distance from the monitor. The design is very inflexible.

Some desktop systems have an input for an auxiliary camera. This must also be manually controlled, but it can be set to image people or objects at varying distances from the desktop unit, giving it somewhat more flexibility than the built-in camera.

Computer-based systems, recommended for a single user, are generally equipped with a very small fixed camera that resides on top of, or next to, the computer monitor. This kind of camera must usually be adjusted manually for focus and brightness. These cameras also typically have fixed focal lengths, requiring the user to sit a predetermined distance from the computer monitor. As one gets closer, the image gets larger and looks unattractive on the display. As one moves back, one's image can be framed better to give a head-and-shoulders image rather than a close-up of the face. This can present a problem for the person whose eyesight or physical capabilities require that he be close to the keyboard, mouse, and monitor. And moving the camera back is often not an option because of limited desk space or the sloping top of the computer monitor. For this reason, many users of computer-based systems upgrade the small, inexpensive camera that comes with their system to a better camera equipped with a *variable focal length* lens. Variable focal length lenses are also called *zoom* lenses.

In cart, or rollabout, systems used in telemedicine, the camera used by a consulting physician is typically set to image the doctor at a fixed distance of 1 to 3 feet from the monitor. It is manually moved up or down and from side to side. The iris (or light setting) is usually adjusted manually. The camera for imaging a patient must be more flexible, however. It must be able to provide full patient images as well as close-ups of affected areas. It is usually mounted on a tripod on castors to enable a full range of positions.

In distance education systems, the instructor stands or sits at an electronically equipped instructional station or podium. The instructor controls a camera, usually mounted on a rear wall, to image himself or herself during instruction. For instructors who pace as they lecture, a *tracking system* can be installed. The instructor wears a small infrared transmitter and wireless microphone. The camera is positioned on a mount that follows the infrared signal. Although this keeps the instructor in view of the remote students, the constant motion can be very distracting to students in a remote location.

8.4   Imaging three or four participants

It is fairly simple for one or two participants seated together to be imaged adequately and to see the monitor of the system they are using. But the challenge of seeing and being seen increases with the number of participants.

Replacing a face-to-face meeting with a videoconference, or an in-person class with distance education, or on-site diagnosis with remote diagnosis requires that each participant be able to derive the same visual information as he or she could obtain if he or she were physically present at the remote site. A video camera can be used to image a single person or a stadium full of spectators. However, if enough detail is to be captured to convey facial expressions or subtle movements, the camera must be focused on *no more than* four people. As the camera angle is widened to image more than four people, the amount of the picture occupied by each person becomes too small to allow the camera to capture details (Figure 8.1). The situation cannot be improved by using a larger display or a better camera because the information was not captured by the video system.

If the application involves four or fewer participants, a single camera will be adequate. The simplest, and least desirable, configuration has a single fixed camera, usually with a variable focal length allowing it to be manually zoomed in or out to image the participants. This design requires participants to be seated in predetermined positions and offers little flexibility.

A better design includes *remote camera control.* In this design, the camera is on a pan and tilt mounting. Participants can reposition the camera up, down, left, and right by using buttons or keys on the system controller. Zooming, focusing, and brightness control are also handled from the system controller. With this design, the camera can be adjusted easily before the session to image the entire group of three

Figure 8.1 Comparison of relative sizes of participants imaged by a single camera.

or four participants. It can be left in this position during the session, or it can be moved during the session to provide close-up images of individuals or pairs of people as the dialog or focus shifts from person to person during the session. Remote camera control is discussed in detail in Chapter 11.

8.5   Imaging more than four participants

When the group size exceeds three or four participants, a single fixed camera is no longer adequate. At this point, the design must include a single movable camera or multiple cameras.

8.5.1   Single movable camera

The single movable camera can also be used for larger groups. However, moving the camera manually from the system controller can disrupt the flow of the meeting. Most large group and some small group systems are therefore equipped with *camera presets*. The system has a small amount of computer memory to store preselected camera positions. Before a session, the user determines desirable camera positions and then presets and stores them in the system's memory. During the session, these stored camera positions can be recalled or even changed. This allows a user to move a camera quickly and via the most direct route to the desired view, thus eliminating a good deal of distracting camera movement. Systems designed for small groups of three or four participants may include up to four programmable positions. Systems designed for larger groups may include up to nine positions. This number of positions is enough to store the positions of individual participants, an overview image of all the active participants, and perhaps an image of a podium or presentation area or an observer seating area.

Some users find it difficult to concentrate on the content of the meeting and also effectively move the camera. Because viewers who are observing the distant video are often more aware of the need to redirect the camera than those in the location where the action is taking place, most large group systems offer *far-end camera control*. A user can take control of the camera in another location and move it. In some systems, they can even activate the distant-end camera presets.

A few systems used in videoconferencing and telemedicine have a feature called *far-end image control*. This allows the viewer to select an alternative video source in systems with multiple cameras. For example, the viewer may select a camera that is focused on an observer seating area, or the viewer may become tired of seeing the image from the document camera and choose instead the camera that images the participants.

For applications where people are always seated in fixed positions, *voice-activated camera switching* offers an alternative approach. This design relies on multiple microphones in fixed positions around a table. A device that senses the relative volume of the audio signal coming from each microphone is used to move the camera in the direction of that microphone in order to image the speaker. Although this is clearly the most user-friendly means of moving the camera, it is seldom used.

One drawback is that the camera movement can be actuated by stray noises such as a cough, a squeaking chair, or rustling papers. This can be overcome. The circuitry can be designed with a delay such that only a sound sustained for more than a second causes the camera to be redirected.

A more serious concern is the frequency with which the camera moves. In any conversation, participants may verbalize assent or make other brief comments. These cause the camera to move. The distant end can experience a dizzying sweep of the room each time members verbalize their reactions. Add to this a delay in switching, and the result is chaotic.

8.5.2 Split screen

The more creative method of capturing a detailed image of more than three participants is known as *split screen* (or sometimes called *continuous presence*). It has been observed that when a camera is focused on three or four seated people, the top quarter of the picture is the background. This is typically the rear wall of the conference room. The lower quarter is the foreground and includes the surface of the conference table. There is no meaningful information in these two portions of the picture (Figure 8.2).

If there are six to eight participants, two cameras can be used to image three or four participants each (Figure 8.3). The two pictures

Figure 8.2   Meaningful area of video image.

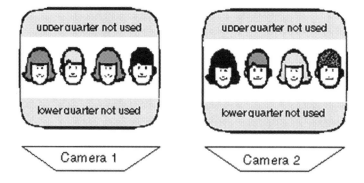

Figure 8.3   Two cameras can be used to image multiple participants.

will have adequate detail. By electronically using only the center half of both camera outputs and stacking them to form one picture, a detailed view of all eight people can be sent to the distant site as a single video image traveling on a single communications channel. At the receiving location, the picture can be displayed in its stacked format (Figure 8.4). This is often called the *Hollywood Squares* format.

An alternative to stacking the pictures is to display them on two side-by-side monitors. This is known as the *Last Supper* format (Figure 8.5). In this format, black is used to fill in the missing portions. The effect is similar to the letter-box format used to display movies on a television screen.

Some large group systems offer the split-screen feature. If it is implemented, there will be two main cameras located near the main

Figure 8.4 Image from two cameras displayed on a single monitor.

Figure 8.5 Images from two cameras displayed on dual monitors.

monitor. The cameras are set up before the session begins and there is no need to move them during the session.

8.5.3 Optimal solution

A combination of camera presets, distant-end camera control, and split screen is often the best solution. When a participant is speaking, presets are used to focus the camera on the speaker. If the person controlling the camera at the speaker's location does not move the camera, the remote location can do it. When there is no speaker at a site, a split-screen picture of all participants is transmitted to the distant end.

8.6   Additional participant cameras

For some applications, it is desirable to have additional cameras that can be used in place of, or in conjunction with, the main camera.

In videoconferencing and other meeting applications, more than one camera may be required for larger groups of people. As just discussed, two cameras support the split-screen option. Additional cameras can be wall mounted or placed on movable tripods to image observation areas.

8.6.1   Presentation cameras

During an in-person meeting, a speaker may make a formal presentation at a podium or a more informal presentation at an easel or whiteboard. This capability can be added to a videoconference installation.

A whiteboard can be mounted on a side wall between the conference table and the system monitors. A podium can be positioned off to the side of the system monitors. In some cases, the main system camera can be directed toward the presenter. But usually the angle of imaging is not ideal and a separate camera is provided. This camera may be a tripod-mounted camera that can be moved into position in front of the presenter. A better solution is a camera either mounted on the wall opposite the presenter or hung from the ceiling. The camera should be positioned so that it can image both the presenter and the writing surface. Since the system monitors are usually off to the presenter's side, in some systems, the presenter is provided with a ceiling-mounted monitor directly in front of him or her so that he or she can observe both local and distant audiences during his or her presentation.

Most installations permit the presenter to move the camera and manage other system functions using the main system controller, if it is portable, or via a handheld auxiliary controller.

8.6.2   Student cameras

Other cameras are required for distance education. In addition to the instructor's camera, an additional camera may be used to provide a wide view of the local students. If a student in the instructor's classroom makes a comment or asks a question, the instructor switches to this camera, so that students in other locations can see the student speaker. In many installations, the instructor can zoom in on the student who is speaking. However, it is not realistic to expect an instructor interacting with students to control local or distant cameras such that he or she will perfectly image every student who is responding to questions. More likely, wide-angle views or views of a portion of the local or distant classroom may be all that can be practically provided, particularly in a lively discussion.

8.6.3 Patient cameras

In remote diagnostic applications, carts or rollabouts with single fixed cameras are deployed. However, the patient is usually imaged with a separate camera mounted on a tripod, placed initially as near to the main monitor as possible. This camera is far more versatile than a cart-mounted camera. It serves as a face-to-face camera during the initial portion of the examination, capturing the image of the patient and on-site health care provider(s). During the examination, the consulting physician can remotely control the pan, tilt, zoom, and focus of this camera. The lens can be set to a wide view to allow the physician to observe the patient walking, for example, or it can be zoomed in for a closer look at a portion of the patient's body. Many systems have a number of camera presets to make it easier for the physician to move the camera.

8.7 Other cameras

It isn't only the people involved in visual telephony who need to be imaged. There are also requirements for imaging other things. Here are some examples of specialized cameras.

8.7.1 Document cameras

Videoconferencing installations often include a separate camera for imaging documents and objects. The most common design is a camera on an adjustable arm. The camera is located above a surface on which the graphic material is placed. Typically, there are lights below the surface that can be turned on when transparencies are used. This device is called a *document camera* (Figure 8.6).

In early systems, a ceiling-mounted camera, sometimes called an *overhead camera,* was positioned above a glass panel built into the conference table. This camera was used instead of a document stand. The permanently mounted camera reduced the flexibility of the system. It also made maintenance a problem. Today many document camera stands are available, and overhead camera designs are rarely encountered.

Most interactive distance education classrooms have a separate camera for the imaging of objects. In some installations, it is the same kind of document stand used in videoconferencing applications. In others, and particularly in those installations that are custom built, a document camera may be mounted in the instructor's podium. In

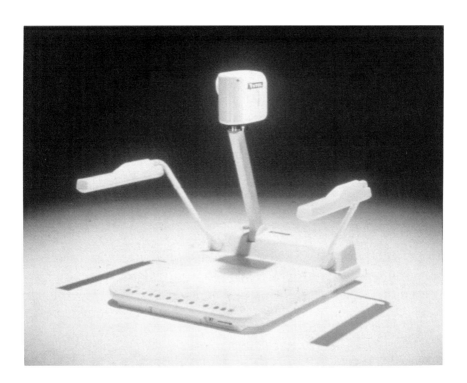

Figure 8.6 Document cameras can image hardcopy, transparencies, and three-dimensional objects. (*Source:* VTEL Corporation, Austin, Texas.)

either case, the ability to display drawings from a textbook, photographs, objects, and other graphics is essential to distance education classrooms.

8.7.2 Diagnostic cameras

A camera such as the document camera used in videoconferencing (see preceding discussion) can be used to provide the consulting physician with a detailed view of some areas of the patient's body. Hands and arms can easily be placed under the lens. The camera body tilts up to image the chest, face, and back. This camera can also be used to image patient charts; telemetry graphs, such as electrocardiogram results; and objects. The close-up camera can be used to image X-rays, but the resolution is not sufficient for an initial reading.

We have all seen doctors using an otoscope. This is the flashlight and lens used to look into the ears and nose and down the throat. In a telemedicine system, the otoscope is equipped with a camera. The local health care worker uses it to show the consulting physician these areas.

Other specialized cameras can be included in a system design. Virtually any device that is used for visual examination can be fitted with a video camera. For example, orthoscopic devices, with a camera attachment, can be used to do knee examinations.

A video camera can be mounted on a microscope so that microscopic analysis can be both displayed on a video monitor and transmitted over a visual telephony link. Some telemedicine installations include a microscope and camera.

8.7.3 Slide-to-video converter

At one time, 35mm slides were a popular visual aid for meetings and instruction. Some people still prefer them. To integrate slides into a visual telephony installation, a slide-to-video converter is used. This product incorporates a video camera into a slide projector. Instead of projecting the slide image on a screen, a video image of the slide is sent to the visual telephony system. The image can be displayed locally on a video monitor and transmitted as a video graphic.

8.7.4 Specialized cameras

Video cameras are used for a wide variety of imaging tasks. Cameras are attached to microscopes, which can then be used in a classroom for teaching biology or for diagnostic and research purposes. Other cameras can be used to display the output of a telescope so that astronomy can be taught from some of the expensive observatories located in remote sites around the world. The list of possibilities is endless. Whatever can be imaged can be transmitted via video telephony.

9

Display Systems

THE LAST CHAPTER dealt with *video input*—capturing useful images with cameras and other imaging instruments. In this chapter we look at *video output,* examining the various options for displaying the images of visual telephony.

9.1 Display devices

Video images can be displayed on a variety of devices. Those most often found in visual telephony are monitors, projectors, and computer screens.

9.1.1 Monitors

The most common means of displaying video is on a *monitor.* A home television set is a type of monitor. A monitor displays images on the surface of a *cathode-ray tube* (CRT). Because the video pictures appear on the CRT, it is sometimes called a *picture tube.*

Video images are made up of individual spots of light aligned along horizontal lines. The spots of light are called *picture elements* or *pixels*. The number of pixels that can be displayed and the number of horizontal lines determine the crispness of the image. This is referred to as the *resolution*. The maximum resolution available from standard television in the United States is 525 pixels on 480 lines or 525 × 480. The typical consumer television has less than this.

The monitors used in visual telephony applications are often high-quality consumer televisions that approach 525 × 480 resolution. These monitors represent a small portion of consumer television sales, but they are the dominant type of display used in commercial applications.

Monitor sizes reflect the diagonal measurement of the CRT surface. The edges of the CRT are rounded. There is distortion in this part of the picture. Monitor manufacturers cover this part of the picture with cabinetry. In the United States, the actual viewing surface is the measurement specified. In Europe, the size of the CRT is specified. As a result, a 29-inch European monitor is the same size as a 27-inch American monitor.

Monitors used in visual telephony range from 10 inches for non-computer desktop units to 45 inches for large group systems. Monitor manufacturers are striving to make even larger products.

9.1.2 Video projectors

Because monitor sizes are limited, *video projection* must be used if a larger display is required. Video projectors can be used in videoconferencing and distance education applications. However, designing a video-conference room or distance education classroom using projection technology is tricky. For best results, the screen onto which the video is projected must be in low light, while keeping the on-camera participants well lighted.

The position of the projector is an important consideration. Highly directional lighting can be used to cast enough light on the participants yet not wash out the image on the screen. However, if the projector is mounted on the ceiling of the conference room, the lights could diffuse the projection before it reaches the screen.

One means of overcoming these problems is to locate the projector behind the display screen. A separate room adjacent to the conference room is used to house the projector and other equipment. The viewing screen is in a wall between the rooms. This is called *rear-screen projection.* Video projectors have a reverse setting for this type of installation. The projector must be far enough from the screen to allow a large display. Rather than allocating a large amount of space to the equipment room, a series of mirrors can be used to provide a larger optical distance than the actual physical distance.

Cabinet projection systems incorporate the projector, mirrors, and rear projection screen in a sealed unit. They have a viewing surface of between 40 and 60 inches. These can be set against a wall or built into a partition. Whether an equipment room and rear projection system or a cabinet projection system is used, this type of display requires a site-built conference room or distance education classroom.

Large-screen displays are desirable in visual telephony for two reasons. First, the size of the audience may dictate a large display so that all attendees are able to see clearly the video being displayed. Second, visual telephony is more realistic when the participants appear to be more life sized.

When viewing a normal-sized monitor at a reasonable viewing distance, the individual pixels are not discernible. However, if a magnifying glass is used to look at the picture on a video monitor or home television, the pixels are clearly visible. Likewise, if one is close to a large display, individual pixels are clearly seen. In some designs, emphasis was placed on having life-sized images close to the local participants. The participants are close enough to see the pixels, or more accurately, the lines separating the picture elements. This detracts from the objective of providing images that appear lifelike.

There is a way of improving the quality of the image. The amount of video information available for display is determined by the amount of information transmitted to the display device. If 480 lines of video are sent to a monitor, that is all the information available for viewing. However, the number of lines can be doubled so that the pixels are less visible. This is called *line doubling.* The picture has the same resolution, but it appears to be denser. Many large-screen monitors and projectors have line-doubling circuitry. External line doublers are available when the monitor does not have such circuitry. When using large display devices, a line doubler should be used.

9.1.3 Flat-screen technology

Videophones and laptop computers use a lightweight thin display. The generic term for this type of technology is *flat screen*. The flat screen on a videophone is 5 or 6 inches diagonally. Large flat-screen displays are available. At present, they are not economically feasible for visual telephony. When they are, rollabout cabinets with flat screen interfaces will provide an alternative to site-built conference rooms and class-rooms.

9.1.4 Computer monitors

Computer monitors display more visual information than video monitors. Instead of 525 pixels on 480 lines, an *SVGA* (super visual graphics array) monitor will display $1,280 \times 1,024$ lines. Desktop computer monitors are 14 to 17 inches. People sit roughly 3 feet from them. If the resolution were not higher than television, the user would see the lines separating the pixels.

This was not the motivation that led computer manufacturers to develop high-resolution displays. If computer monitors had the same resolution as consumer television sets, printed text would need to be larger for the user to be able to read it. A typewritten page, which can have as many as 80 characters on a line, would not fit on a television, which can only display 40 characters per line. Likewise, all graphics would need to be larger and less dense. A spreadsheet would probably be impossible. Hence, there was a need for computer monitors to have higher resolution in order to display a printed page.

When a computer monitor is used to display visual telephony, the video resolution is limited to what has been transmitted. Visual telephony does not have resolution near that of SVGA. The video can be placed in a window and the rest of the monitor screen can be used for text or graphics. In the small window, the video appears sharp and lifelike. This display allows users to collaborate on the information displayed on their screen, while still maintaining visual contact with each other.

Most computer-based products allow the user to expand the incoming video to fill the entire screen. A self-view image may be placed in a small window with this display configuration. The product has essentially been line doubled to allow the lower resolution video to be displayed on the SVGA monitor. The resulting image is generally very appealing.

Some group applications of visual telephony are being built around a computer-based product. However, a large SVGA monitor is used to make the system acceptable to larger groups of users. As noted in Chapter 3, with the addition of a 27-inch SVGA monitor, variable focal length camera, and infrared keyboard, a desktop system can be converted into a small group system.

9.2 Video formats

In the United States, broadcast television resolution is 525 × 480. This standard is referred to as *NTSC*, derived from the National Television Standards Committee that adopted the standard in 1954. This standard is used throughout North America, Japan, and elsewhere. There are other standards. In Europe, the television image is 625 × 576. Two European standards use this format. Most of Europe uses *PAL* (Phased Alternating Line). France and some countries in eastern Europe, Asia, and Africa use *SECAM* (Sequentiel Couleur Avec Memoire, or Sequential Color with Memory).

NTSC and these other formats use a picture that has a length that is one-third more than the height. Another way of stating this is that the *aspect ratio* is 4:3 (Figure 9.1). This format was selected because it was the format used in motion pictures at the time.

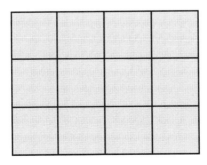

Figure 9.1   The 4:3 aspect ratio of NTSC, PAL, and SECAM television displays.

Later, the film industry changed to a more panoramic format. Now, films that are shown on television or converted to videotape must be digitally modified to fit on television screens.

A new *high-resolution* video standard has been approved for the United States. The resolution will be 1,000 × 1,000 and the new aspect ratio will be 16:9 (Figure 9.2). This is similar to motion picture film. Today, monitors and cameras that operate using the new standard are too expensive for use in visual telephony applications. This will change shortly, as products enter mass production. High-resolution systems will begin to edge out conventional systems within the next 5 years. High-resolution flat-screen systems will add a whole new set of options for visual telephony designs.

9.3 Monitor location

The location of the monitor(s) is critical, both for ensuring that users can comfortably see the incoming image and for obtaining the most natural outgoing image of that viewer.

9.3.1 Videoconferencing

To provide a sense of eye-to-eye contact, the cameras should be located where the user's eyes would naturally rest, and that is usually on the monitor that provides video from the distant location. For videoconferencing applications, then, monitors are usually located in front of a conference table.

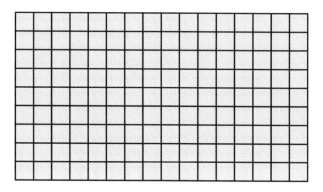

Figure 9.2 The 16:9 aspect ratio of high-resolution television.

9.3.2 Distance education

Displays for the instructor

For distance education applications, the situation is somewhat different. The instructor must be able to see the students at the remote site(s). If there is also a local class, the instructor must be able to see—and maintain eye contact with—both the local and remote classes. In some designs, a monitor is built into the instructor's station. When the instructor is teaching to both local and remote students, the monitor should be situated so that the teacher can both observe the remote class and look over the monitor at the local students. Otherwise, the teacher will lose visual contact with the distant class. Another option is to install the monitor on the rear wall. The instructor camera is placed just above it. The instructor can easily view both local and remote classes and both have the benefit of eye contact.

In addition to having a monitor to show the distant students, the instructor should have a monitor that displays the video being transmitted. This is usually built into the instructor's podium or workstation. In many systems, this monitor can also be used to preview the output of any camera before it is selected for transmission. Other systems add a dedicated preview monitor, built into the podium or workstation.

Displays for the students

Distant students must see both the instructor and any teaching aids (charts, videotapes, computer-generated graphics, annotation) the instructor transmits. Local students don't need a monitor to see the instructor, but they must be able to see the teaching aids. In some installations, they can also see remote students.

Display devices for students may be one or two large monitors located in the cabinetry of the instructor's station, below the instructor's work surface, or that are mounted on the wall, near the instructor.

Large projection displays can also be used for interactive annotation, which is a direct replacement for the traditional blackboard. They can be used for displaying a videotape or video transmitted from the remote classroom. Large projection devices are located near the instructor.

9.3.3 Health care

Most remote diagnostic systems have one main monitor. The consulting physician sees the patient and the health care provider on this monitor, which is used for viewing all of the video images sent from the patient site.

At the patient site, the main monitor displays the image of the consulting physician. Because the health care worker at the patient site may be asked to assist in moving or focusing one of the cameras, a monitor that shows the camera output is required. In some designs, the main monitor is switched to local display for this purpose. In other systems, a smaller preview monitor is provided. The smaller image is less intimidating to patients who are self-conscious about seeing their own images. In other designs, a picture-in-picture on the main monitor is used. In all cases, the self-view image can be switched off at the patient's request.

9.4 Typical monitor configurations

Several configurations of monitors are available to accommodate the various kinds of visual telephony systems. The configuration that best suits the application is the one that should be selected.

9.4.1 Large group rollabout systems and custom rooms

During in-person meetings, people often use visual aids to assist in presenting information. Charts, graphs, objects, photographs, and other presentation aids are typical. In an in-person meeting, people can shift their gaze from the graphic to the presenter and back again, at will. Large group systems have two monitors and the ability to store a still video image, which can be displayed on the second monitor. Participants can look at either the graphic or the distant participants, just as they can during an in-person meeting.

Systems with split screen use the second monitor to show the two halves of the incoming picture, side-by-side. When a graphic is transmitted, the two halves of the split screen picture are stacked on one monitor and the graphic is displayed on the other (Figure 9.3). The viewer can switch back and forth between this display configuration

Figure 9.3 Split-screen configuration during graphic presentation.

and the split screen pictures of participants on the two monitors, with no graphic. Since the graphic is in memory, it can be retrieved again easily.

Large group systems can be used to transmit prerecorded material such as videotapes (Figure 9.4). The second monitor is used to display the tape when played locally. The other monitor displays the motion video from the distant site. This could be a stacked image or the output of a single camera. Custom rooms may use a third monitor to display the videotape, leaving the main monitors for the split screen images from the remote site.

The system must provide a means for the user to see the signal that is being transmitted to the distant site. In some systems, the outgoing image can be displayed on one monitor while setting camera

Figure 9.4 Split-screen configuration during videotape presentation.

presets. Because a large self-view is distracting and because the second monitor is better used for graphics or split screen of the distant site, the full-screen self-view is generally switched off during a meeting.

A small self-view is useful in order to be certain of what is being transmitted. Most systems place a small picture of the outgoing signal in a window on the main monitor. This feature is known as *picture-in-picture*. The picture-in-picture can be moved if it is blocking an important portion of the incoming picture.

Some systems have a dedicated *self-view monitor* for the display of the outgoing signal. It can be the same size as the main monitor(s); however, it is usually a smaller monitor placed to the side of the main monitor(s) or on the conference table. Custom-built rooms that have a third monitor on the side of the main monitor(s) may use it as both a self-view monitor and a monitor to display graphics or a videotape.

9.4.2 Summary of monitor configurations

There are a number of ways to use video monitors in large group systems. Figures 9.5 through 9.12 summarize the possibilities.

9.4.3 Small group rollabout systems

The standard small group system has one main monitor. It is used primarily for viewing the motion video from the distant site. Most small group systems use picture-in-picture for self-view.

All small group systems have the ability to store a still image graphic and display it on the main monitor. In some systems, when a graphic is being displayed on the main monitor, the self-view picture-

Figure 9.5 Two main monitors showing split-screen (two-camera) destacked view and picture-in-picture self-view.

Figure 9.6 Two main monitors showing split-screen (two-camera) stacked view, graphics display, and picture-in-picture self-view.

Figure 9.7 Two main monitors showing standard (one-camera) view, graphics display, and picture-in-picture self-view.

Figure 9.8 Two main monitors showing split-screen (two-camera) destacked view and an additional self-view monitor.

in-picture can be replaced with a small image of the motion video from the distant site. These systems also allow the graphic to be shown in the picture-in-picture window when the main monitor is displaying the incoming motion video. Other systems offer an optional second monitor for graphics. Analyze the cost of this option before buying. At times, adding features such as a second monitor to a small group system can be more expensive than buying a large group system that offers a second monitor as a standard feature.

Figure 9.9   Two main monitors showing split-screen (two-camera) stacked view and graphics display, along with an additional self-view monitor.

Figure 9.10   Two main monitors showing standard (one-camera) view and graphics display, along with an additional self-view monitor.

Figure 9.11   Three main monitors showing split-screen (two-camera) destacked view, graphics display, and picture-in-picture self-view.

9.4.4   Desktop systems

Desktop systems have a built-in monitor that is usually no larger than 10 inches in diameter. A single user can comfortably use the system while sitting within 3 feet of the monitor. Two people can use a desktop system in this manner, but if there will be a number of occasions when two people use the system for long meetings, a larger monitor is advisable. Some desktop systems have a video output so

Figure 9.12 Three main monitors showing split-screen (two-camera) destacked view, graphics display, and an additional self-view monitor.

that a larger monitor can be added to the system. If not, a small group system, rather than a desktop system, is advisable. Small group systems can have monitors as large as 27 inches. Participants can sit further away from the monitor, they can see better, and there can be more space between them.

9.4.5 Computer-based systems

When a number of people will be collaborating with each other using their computers as interactive workstations, computer-based systems are the right products. As noted earlier, the incoming video is generally displayed in a window on the computer screen. The user can vary the size of the window. Self-view replaces the incoming video. Therefore, self-view is only used to verify camera settings and generally not used during a session. However, when there is no computer-generated material on the screen, the incoming video can be displayed on the entire screen with a picture-in-picture window for self-view.

9.5 Size of video displays

Thus far, we have only noted that the different types of systems have different sized monitors. Desktop systems include a small monitor under 10 inches. Computer-based systems display the incoming video in a portion of the computer's monitor. Group systems can include monitors that range from 19 to 45 inches. How does one determine the correct display size for an application?

The correct monitor size can be calculated based on the number of participants that will be viewing the display. (Remember that monitor sizes are all measured as a diagonal from one corner to the opposite corner.) A detailed description of these calculations and a tutorial on how to use them are given in Appendix C. Table 9.1 shows the monitor sizes that should be installed based on the number of participants.

A system for more than four participants must include capabilities for large-screen projection, split screen, or moving the camera. Generally, large-screen projection is not used, and all active participants are kept within the 10.5-feet requirement, imaged with a movable camera or a split screen system. Observers should also be approximately 10.5 feet from a monitor. In some cases, this requires that additional monitors be located in the observation area. See Chapter 8 for a discussion of split-screen camera configurations.

This chapter has discussed considerations for selecting and placing display devices. *There is no simple formula that suits all applications.* Choose the best approach for your application and your facility. Consult Chapter 13, which covers facility considerations, before choosing a device.

Table 9.1
Monitor Sizes and Distance of Participants

Number of Participants	Distance From Display (feet)	Screen Size (inches)
1	2.5	5
2	5.0	13
3	8.0	21
4	10.5	27

10

Audio

Although the essence of visual telephony may seem to be the video elements (camera and display), it is the audio that is of most importance. People will tolerate a less-than-perfect video image, but will not be able to communicate if the audio is unintelligible.

Audio systems for visual telephony installations include speakers, microphones, echo-cancellation electronics, and audio add-on features. How well they function is related in part to the environment in which they are installed.

10.1 Speakers

Large group rollabout systems typically have two speakers built into the system's cabinet. Custom-built rooms may have several. They are typically located in the ceiling.

Small group systems have one speaker, usually built into the monitor.

In telemedicine systems, there is usually only one speaker, located in the telemedicine cart. Because the level of the received audio can affect patient confidentiality, some additional features are necessary. Some designs include a headset for the consulting physician. The headset prevents incoming audio from being heard by people nearby. Other designs have a speaker that can be removed from the cart and placed near the patient. A level adjustment on the speaker gives the patient the opportunity to set the volume to a comfortable listening level.

Computer-based systems use the speakers that are part of a computer's multimedia system. Headsets or speakerphones are other options.

10.2 Microphones

Many different types of microphones are available for visual telephony applications. The key issues are frequency response, pick-up area, and aesthetics and practicality.

10.2.1 Frequency response

Inexpensive microphones convert only a narrow range of frequencies into electronic signals. A telephone is a good example. When people call into a television or radio show, the difference between the broadcast audio and the telephone audio is readily perceptible. The telephone transmits fewer of the frequencies present in the human voice. In technical terms, broadcast television audio ranges from 50 Hz to 15 kHz, whereas basic telephony is between 50 Hz and 3.5 kHz. (Hz is the abbreviation for hertz, a unit that indicates cycles per second of a frequency; a kilohertz is 1,000 Hz.)

Group videoconferencing systems, distance education systems, and telemedicine equipment are capable of transmitting audio in the range of 50 Hz to 7 kHz. The microphone selected should be capable of operating in at least the 7-kHz range without distortion.

Desktop systems have audio frequency that is in the same range as telephones. The microphones used for this application need not be of the same quality as those used for other visual telephony applications.

10.2.2 Pick-up area

Most of us have seen performers holding a microphone. Performers frequently hold the microphone very close to their mouths. This keeps the microphone from picking up any sounds other than the performer's voice. If the microphone is farther from the person singing or speaking, sounds from a broader area are detected. The microphone's pick-up area is shaped like an ice cream cone, with the pointed tip located at the microphone. The design of the microphone determines the pick-up angle, or how broad the cone is.

Tabletop microphones used in visual telephony have a pick-up angle of 120 degrees. By locating the microphone on the conference table facing the participants and about 2.5 to 3 feet from their seating positions, the microphone can pick up the voices of three to four people. The area to the sides and behind the microphone is outside the cone pickup area and should detect as little as possible. In videoconferencing, this is where the videoconference system is located. Neither the distant audio coming through the speakers nor noise generated by the equipment should be transmitted.

Ceiling microphones should be selected with a great deal of care. If the pick-up area includes noise-generating equipment or air conditioning vents, the ambient noise can be very distracting. In a distance education design, a microphone that takes in a large number of students is certain to add a great deal of stray noise to the audio.

The microphones used in telemedicine applications should have a narrow pick-up angle. These systems are often used in examining rooms that cannot be treated with sound-absorbing material.

The microphones used for desktop systems should also have a narrow pick-up. These systems will be used in an office environment. Many offices are not acoustically good for electronic conferencing.

10.2.3 Aesthetics and practicality

The handheld microphone favored by performers is not suitable for visual telephony. This kind of microphone is either intimidating to some or a distraction to others who spend too much of the conference adjusting it. In contrast, the microphones used in visual telephony are designed to be inconspicuous. That is a mixed blessing. While some participants are surprised to learn that the small rectangle on the table is a microphone, others are so unaware of this fact that they cover it with meeting materials. Many audio problems are solved by simply moving documents off to the side. A good rule of thumb is that the microphone should be on the opposite side of the conference table from the participants. If the table is large enough, the participants will not find it convenient to put anything over the microphone.

In some designs, participants are given lapel (lavaliere) microphones. These tend to be more trouble than they are worth. If they are wireless, participants must wear transmitters. If they connect to the conference table, accidents occur when participants forget they are wearing the microphone and get up quickly. Finally, the attire people wear to a videoconference may not have a convenient place to hang a microphone. Conferences have been disrupted by microphones falling off or clunking against jewelry. Tabletop or ceiling microphones should be used for participants seated at the conference table.

In some designs, ceiling-mounted microphones are used to replace the tabletop and wireless microphones. Ceiling mounting removes the microphone from the user's reach. This can be beneficial, because in this position they are less intimidating, not able to be covered up with papers unintentionally, and not able to be manipulated by a user who is trying to take control of the meeting. Mounting the microphones in the ceiling removes the exposed wires and eliminates the hazard of tripping over microphone wires run over the floor when the wires cannot be installed in a floor duct. Ceiling-mounted microphones reduce the portability of rollout systems, as does ducting the microphone wires.

People making a presentation at a podium can be heard with a podium microphone. A low-profile tabletop microphone can also be used here. A ceiling- or wall-mounted microphone should be provided for people working with a presentation board.

Wireless microphones can be available for the occasional use of observers. However, care should be taken to turn them off when not in use.

The microphones used in telemedicine and desktop conferencing are typically small devices on a flexible stand or arm. They should be as inconspicuous as possible.

10.3   Number of microphones

The number of microphones depends on the number of participants and on the design of the facility.

10.3.1   Conferencing facilities

In custom-built rooms and large group systems for meeting applications, there is one microphone for every two or three participant seating positions. If a podium has been provided for a presenter, there will be a microphone located there. There may be a microphone located near a whiteboard, if one is included in the room design. In some systems this is a wireless lapel microphone; in others it is built into the room. There may also be one or more microphones located in an observation area. These can also be wireless.

In contrast to the multiple microphones found in custom rooms and large group systems, most small group systems have one microphone. Since small group systems are designed for four or fewer people, one microphone is enough for the participants seated at a conference table. However, if a whiteboard or observation area is required, or if the number of participants increases, additional microphones must be added.

Once a second microphone is added, it must be connected using a device known as a *microphone mixer*. Large group systems have a microphone mixer as a standard part of the package. Small group systems typically do not. A microphone mixer can combine the inputs from any number of microphones. The mixers used in visual telephony combine either four or six inputs. The second microphone added to a small group system becomes very expensive, because the price includes the cost of the microphone mixer as well as the microphone. Additional microphones, beyond the second one, are the same price as additional microphones added to a large group system.

10.3.2   Electronic classrooms

The most difficult part of designing an electronic classroom is the audio. There must be a wired microphone for the instructor at every location the instructor uses, or the instructor must be provided with a wireless lapel microphone. The audio subsystem should also be able to pick up the local and remote students when they are speaking normally. Microphones are generally shared by every two or three students.

Controlling noise is of paramount importance, especially in distance education applications, given the number of participants and the nature of students. Some designs limit the noise entering the audio subsystem by controlling when the student microphones are active. In basic installations, the student must push a button on the microphone to be heard. This can work, but the students at the origination site often forget to activate their microphone. By speaking naturally, they are heard by the instructor and their local classmates, who may not realize that the distant students have been excluded.

A variant on this approach relies on the instructor activating the microphones. There is an audio subsystem available that incorporates an infrared controller into the microphone design. Each student microphone is equipped with a light and an infrared pick-up. For example, if the instructor wants to include comments from the student in the second row, third seat, the instructor enters ''2-3'' on an infrared keypad. The student's microphone light comes on, indicating that this microphone is activated.

The instructor activates the microphones in the distant classroom the same way. Instead of an infrared signal, a data signal is sent to the infrared sender in the distant classroom. The data are transmitted along with the audio and video. A telephone in the front of the room is available for students to get the instructor's attention without being recognized. This is for emergencies and equipment failures.

Directional ceiling microphones can also be used in classroom applications. The instructor's microphone has a broad pick-up area that covers the entire front of the classroom. The student microphones are more directional to avoid picking up stray sounds.

10.3.3 Telemedicine systems

Telemedicine systems contain a microphone located near the main monitor to pick up the voice of the consulting physician; a headset is optional.

A wired microphone is available for the patient and local health care provider. It is usually stored on the telemedicine cart, in clear view.

10.4 Ambient noise

Microphones pick up more than just the voices of a classroom, meeting room, courtroom, or examining room. Stray sounds like paper rustling and someone moving a chair are also transmitted to the remote site. The more microphones there are in a room, the more likely it is that stray noises will be transmitted along with the voices.

When we are listening to music or concentrating on someone speaking in the same room, we have the ability to filter out noise mentally. This is because the noise is coming from a different place than the music or speech. Our brain can focus on the sounds reaching it from one source and push noises from another location into the background. However, when stray noises are added into the audio transmitted from a visual telephony site, the participants at the receiving location hear it coming from the same speakers as the voices they are attempting to hear. Their brains are no longer able distinguish the sound from the noise and the intended sources. Incidental noise is very distracting.

Chapter 13 discusses how to select a site and treat it to reduce ambient noise. In addition to the characteristics of the room itself, the selection and location of microphones play a major role in reducing the stray noise that is allowed into the conference audio. This part of the system deserves serious consideration.

10.5 Echo cancellation

Just as the video operates in more than one direction and is interactive, the audio should be capable of simultaneous transmission in all directions.

When a person speaks, his or her voice is transmitted to the speakers at the distant site. There, the sounds coming from those speakers are captured by the microphones. If that incoming sound is allowed to be retransmitted, the person talking will hear his or her own voice, delayed slightly by the time it takes for the round-trip transmission. The result is an echo. The potential for echo occurs whenever there are open microphones and speakers. Distance education classrooms are particularly vulnerable.

Most of us are familiar with speakerphones. The approach used to overcome echo in these low-cost devices is known as *gating* (or *half-duplex*) circuitry. Both the incoming and outgoing audio are routed through the gating circuitry. The gate will allow audio to pass in only one direction. The loudest signal is allowed to go through. This is presumed to be the voice of the person speaking.

Speakerphones are not intelligent, however. When there is a loud background noise, such as a passing truck, the incoming signal is interrupted, causing a breakup in the transmission. In a lively dialog, the give-and-take of comments can also cause interruptions. This effect is called *clipping.* It can be very distracting.

A better approach is to use a device known as an *echo canceller.* An echo canceller converts the incoming audio into a digital signal that computer circuitry can store in memory. This audio is allowed to pass through to the speakers. The audio that is picked up by the microphones includes both the voice of the distant person speaking and any comments made by local participants. All of this is routed to the echo canceller, where it is also digitized. The incoming and outgoing signals are compared, and the incoming portion is electronically removed from the signal transmitted. Only the sounds made locally are sent to the distant site. Unlike gating, both sides can speak simultaneously and still be heard. There is no clipping. (The use of an echo canceller is often referred to as *full-duplex* audio.)

Audio is affected by the acoustics of the room, which will be different from session to session based on the number of people in the room and the arrangement of equipment. Echo cancellers train themselves on the session in the first few minutes of being activated by testing the acoustics and adjusting accordingly to provide the best possible audio for a particular session.

Echo cancellation is usually standard on large group systems and is provided by most manufacturers of small group systems. However, this is an important consideration and should be verified before buying any system.

10.6   Telephone add-on

During a meeting, it may be necessary to call someone for information or clarification or to schedule a facility or appointment. The call can be added into the conference audio so that all participants can both hear from and speak to the person called. In distance education, the ability to bring in an expert from industry, government, and other educational institutions as a guest lecturer via a telephone connection can be very valuable. Likewise, connecting immediately with an expert to provide additional information to assist in remote diagnosis can be a very useful feature.

In some instances, a conference bridge is added to the audio system so that people at remote sites without visual telephony can be audio-only participants.

The ability to add one or more callers to a visual telephony session is known as *telephone add-on.*

10.7   Improving the audio environment

Today's audio electronics function well in most acoustic environments. However, attention to sound treatment can improve the performance of any audio system.

Conference rooms usually have carpeted floors and chairs covered in cloth and stuffed with foam. These soft surfaces are desirable for audio systems.

In contrast, examination rooms seldom have carpeting and the chairs are plastic. The harsh conditions make it easier to maintain a clean environment, but they also make the room subject to echo.

Likewise, in distance education, rooms tend to include hard surfaces (floors, walls, windows, furniture) and noisy participants. Here, the classroom must be treated to counteract the environment and to reduce the effect of ambient noise, such as traffic on the street outside or human traffic on the playgrounds and in the corridors.

Acoustic treatment need not add a great deal of expense. As discussed in Chapter 13, there are inexpensive ways to dampen ambient noise and absorb extraneous sound. Any expense, however, to improve the audio part of visual telephony is usually money well spent. People will put up with poor image quality, but they won't tolerate poor audio quality.

11

Control Systems

A MERE COLLECTION OF cameras, monitors, speakers, micro-phones, and other equipment won't make for a great visual telephony experience, even if all the equipment is top-notch. In order for the experience to be worthwhile, the user must be able to control the components in a natural and useful way.

11.1 Design objectives

Visual telephony is used to accomplish a communications objective—instructing, diagnosing, testifying, giving or getting information, for example. In order for the user to be able to accomplish the communications objective, he or she must be free from complex or time-consuming manual system operations. A user-friendly control system is therefore essential to the successful operation of a visual telephony system.

The user interacts with the visual telephony system through the control system. The more features a system provides, the more important the design of the controller. A well-designed control system should:

- Incorporate all the equipment in the system, allowing the user to manage the equipment from a single point;

- Be easy to understand and use;

- Be reliable.

11.2 Types of systems

A number of types of user interfaces are available for visual telephony control systems:

- Infrared handset interacting with on-screen icons and menus;

- Tabletop systems, including a tablet that interacts with on-screen icons and menus and touch-sensitive screens;

- Mouse and keyboard interaction with on-screen icons and menus.

11.2.1 Infrared handsets

An infrared controller is a small handheld device similar to those used to control home televisions or videocassette recorders. Activating the controller causes menus to appear on the display device. The user can activate a system capability by pushing the correct buttons on the handheld unit.

Infrared handsets are compact and easily passed from participant to participant. This is a strong plus in an organization that is striving to establish an open dialog among group participants.

Another benefit of infrared technology is the fact that multiple controllers can be used. A person very familiar with the system may sit off-camera and assist the person running the meeting. This is particularly helpful when users are being introduced to visual telephony.

On the negative side, the amount of physical space on a handset is limited, which in turn limits the number of dedicated controls and the size of writing or icons used to identify these controls. The use of an infrared controller therefore tends to be less intuitive. Another negative aspect is that the appearance of on-screen menus can be disruptive when activated during a visual telephony session.

Handsets are used in custom-built rooms, and in large and small group systems. It is a good idea to have several handsets available in case the main unit is misplaced. And don't forget about fresh batteries! Controllers can begin to act erratically when batteries are low.

11.2.2 Tabletop systems

Tabletop controllers are easier to use than handsets. They are larger, enabling more physical controls and more identifying marks for these controls. Therefore, they can be more intuitive.

Tabletop controllers are connected to the system by a cable. Being larger than infrared handsets and restricted by the cable, they are less easily passed among participants. Also, the cable must run through a duct, along a wall, or be covered to reduce the risk of someone tripping on it.

There are two types of tabletop controllers. One is used to interact with on-screen icons, much as infrared handsets. Just as with an infrared controller, the on-screen icons and menus can be distracting. In some designs, the tablet has a pen device that is used in conjunction with on-screen icons. The icons are also printed on the control tablet.

The other type has a touch-sensitive screen on which icons and menus appear. With this type of controller, there is no need for on-screen icons. The management information appears on the touch-sensitive screen, and only the control operator is aware of it. However, most systems do display a few critical icons or messages on the viewing screen. The most common is the *microphone muted* indication.

These devices generally have multiple pages so that the user can easily step through the commands to activate the desired capability. Some systems are so well designed that there is no need for user training beyond identifying the system capabilities.

Tabletop systems are found in custom-built sites, and in large and small group systems. These systems are often used in telemedicine, with the touch-sensitive screen or tablet built into the telemedicine cart. The controller should be intuitively obvious to people merely acquainted with the system. Both the consulting physician and the health care worker assisting the patient should be able to use the system without needing to consult manuals or flowcharts.

11.2.3 Mouse and keyboard

Most people familiar with Apple computers or PCs running Windows have become accustomed to using a mouse and on-screen icons. Computer-based systems, of course, use this kind of control system, but so do many other visual telephony systems. Most actions are performed with the mouse. Dialing, entering directory information, and on-screen messaging functions are done with the keyboard.

11.3   How much control? Who controls?

At the beginning of this chapter, control system objectives were said to be completeness, ease of use, and reliability. The challenge of selecting or designing a control system is made more complex when there are multiple users, each with a different set of requirements.

11.3.1 Videoconferencing

For example, in videoconferencing, there may be users with various levels of training and experience and a range of meeting requirements from simple discussion to elaborate presentations and group interaction. Many systems have more than one type of controller. Individual users can select the type of controller that is most appropriate for themselves and their meetings.

One popular mixed approach uses a tabletop touch-sensitive screen to control most of the meeting functions. However, each participant has a small infrared device that resembles the transmitter of an automobile remote alarm. The individual controller has only two buttons. One activates the camera preset that is focused on that seating position. The other button returns the camera to the group picture preset.

11.3.2 Distance education

In distance education applications, there are three kinds of users: instructors, students, and technicians. Virtually all distance education origination sites have a controller built into the instructor's station or podium. Most sites have an auxiliary controller either in the rear of the classroom or in an adjacent room so that instructors or guests not familiar with the system can be assisted. Some of the older systems were designed to be operated only by an assistant, but these are not commonly implemented today.

The instructor's control system should have these capabilities:

- Select the video source sent to the remote classroom(s).

- Pan, tilt, and zoom the cameras in the origination classroom.

- Adjust the receive audio level.

- Select the image transmitted from the remote classroom(s).

- Pan, tilt, and zoom the cameras in the remote classroom(s).

Receive-only sites also need some control functions. Students need to be able to adjust the receive volume within a range that is limited to about 6 dB. This allows some adjustment to obtain a comfortable listening level, without creating a potential problem. Students should not be able to raise the volume enough to override the echo canceller, nor should they be able to turn the sound off.

The students will need to be able to zoom and focus the document camera in order to show the instructor their contributions. Students do not, however, need to have access to pan, tilt, and zoom controls for the main camera. These functions are better controlled by the instructor. The selection of the camera as the transmission source (main camera or document camera) also can be left to the instructor. These controls, however, should be available for local technicians to use for maintenance purposes. Many receive-only distance education facilities have two controllers. One is a minimum-function controller for the students, and the other is a full-function controller for the technical staff.

Classrooms that are both transmit and receive facilities must have a full set of controls for the instructor to use. These controls should be inactive when the site is used as a receive classroom. This can be done electronically, or by placing the controls in the podium or workstation so that they can be secured when not in use.

11.3.3 Telemedicine

In remote diagnosis applications of telemedicine, the users are the consulting physician and the health care worker in a remote site with the patient. The consulting physician should be able to control all of the cameras at the patient location. This includes switching to a different video source and, if the camera is so equipped, panning, tilting, zooming, and focusing the image. Because medical experts may not be experienced users of these systems, the control system should be intuitive.

The health care provider assisting the patient should also be able to control the system. The flow of information is enhanced when this person can readily control the visual images transmitted to the expert.

Desktop systems installed in patient homes must include controls that are extremely simple for a patient to operate. They may have minimal functions such as emergency call initiation, receive volume control, and microphone and video privacy. Other capabilities can be activated by the monitoring site.

11.3.4 Judicial applications

The typical courtroom system is controlled by a court security officer. The systems tend to have many capabilities so that they are flexible enough to handle a variety of situations. The judge must also have control over some of the capabilities. This is typically limited to selecting the image that is on the judge's monitor and to muting the bench microphone(s).

At the detention center, there are no prisoner-operated controls. The system will have a means for a technician to control the system during setup and testing. Prison systems used for parole hearings are the same.

Systems used for taking depositions and victim support are identical to videoconferencing systems. They have the same control options.

11.4 Selecting the right controller

Most people test drive a car before buying it. People should do the same when buying a visual telephony system. If the vendor takes charge of the controller and demonstrates all the systems features, beware. The system may be very difficult to control. Over the course of the system's useful life, training costs or the labor costs of having an operator run a complex controller can add up. It is well worthwhile to verify the ease of use before buying.

12

Ancillary Equipment

EVERY APPLICATION OF visual telephony is different, so it is not surprising to learn that additional equipment is required to support specific communications requirements.

The major difference between large group systems and smaller systems is the ability to integrate additional pieces of audio/visual equipment into the system. Large group systems are designed with additional audio and video input and output connectors to accommodate a wide range of ancillary equipment such as video printers, display equipment for computer-generated graphics, videocassette recorders, CD players, and fax machines. In contrast, small group systems are seldom designed to support more than a few pieces of ancillary equipment. For example, in most small group designs, there are only three video inputs that might be used for a main camera, document camera, and presenter camera.

A wide range of ancillary equipment is frequently installed in visual telephony systems. The following pages describe some of the most common equipment.

12.1   Electronic annotation

The ability to write over or highlight a video image is a desirable feature for presentations, diagnosis, and instruction.

The user draws on a tablet or pressure-sensitive screen with a stylus or magnetic pen. Tablets may be located at a conference table, on an instructor's workstation, at a presentation podium, or on a telemedicine cart. A pressure-sensitive screen may be a monitor built into the instructor's station or podium. One product incorporates a cabinet projection system that can display images at up to SVGA resolution, a touch-sensitive screen, and several multicolored electronic pens. (See Chapter 9 for a discussion of SVGA resolution.) Both the local screen and remote screens display what is written by users at any location.

For distance education, electronic annotation is the twenty-first century equivalent of the blackboard. The instructor can write, in various colors, on a blank screen. In addition, the instructor can annotate over the images of objects or text. Receiving classrooms can also have annotation devices installed so students can interact graphically with the instructor and other students. Best of all, there is no chalk dust!

12.2   Scanners

In telemedicine, a scanner may be included with a visual telephony system so that X-rays and other detailed material can be scanned into digital, rather than video, format. The images can be sent, as computer-to-computer data, to the system at the consulting physician's site. The data are carried over the network along with the digital voice and video. The scanned images can be viewed on the main monitor (a scan converter may be required if that monitor cannot display computer resolution images). This method of transferring X-ray images produces high-quality pictures, but it does not meet the requirements for teleradiology. It does allow the physician to review an X-ray that has been previously read by a radiologist. See Chapter 5 for a discussion of the resolution requirements for teleradiology.

Scanners are also useful in preparing materials for incorporation into computer-generated graphics for instruction or presentations.

12.3 Computer-generated graphics

Many people use computers to generate presentation graphics and teaching materials. You can, of course, print paper copies for handouts and produce traditional overhead transparencies or 35mm slides. A more efficient use of computers is to display directly the output shown on the computer's screen. In this way, a presentation or instructional session can be updated regularly, possibly even during the session. It can also be electronically annotated.

In distance education, the instructor can design a block of instruction that is visually rich with text, computer-generated graphics, photographs, and other images scanned into digital files. Animation can be used, for example, to show the flow of a process or to build a mathematical equation to illustrate scientific or business principles. Computers can also be used to analyze student responses collected with automatic response systems and graphs of the results can be displayed instantly. Student work submitted on diskette or by modem can be shared with a class. The increasing integration of computers into classrooms radically improves the quality of local instruction and does the same for students at a distance.

How does one display these computer-generated graphics both locally and to a distant site? There are actually several possibilities.

12.3.1 Traditional hardcopy via document stand

One could print the color charts and place them on a document stand to be imaged by the document camera. However, this is a time-consuming and expensive way to produce an effective presentation.

12.3.2 Computer images via document stand

The second option is to avoid the step of making paper copies and instead feed the output of the computer directly into a display device. In this way, an excellent presentation can be produced without paper.

The information content can be changed right up to the minute of the presentation. In fact, it can be changed during the presentation.

Because the output of a computer (VGA or SVGA) is different from the output of a video camera (NTSC or PAL), it is not possible to transmit computer images to remote sites in the same way video images are transmitted. This second option, therefore, requires some conversion.

One means of integrating computer-generated material into a video-conference is by converting the computer output to video. For in-person presentations, professional presenters commonly use a flat screen that is designed to be placed on an overhead projector, which provides the light source and projects the computer images for a local audience.

Pointing a camera at the projection in order to transmit it to a distant location would yield a very poor quality image. This practice should be avoided. Instead, use the flat-screen device, but rather than placing it on an overhead projector, place it on the document stand, which provides the light source. The document stand's camera produces a video image of the computer graphic much more directly. The image will be seen locally as well as at the remote sites(s).

12.3.3 Scan conversion

A better way of converting the signal requires the use of a *scan converter*. A scan converter changes the computer output into video that can be selected for transmission just as any other alternative source, such as a document camera or presenter camera.

12.3.4 Serial port connections

The last option is more cumbersome. The serial port of a computer can be connected to the data port of a videoconferencing codec in the same way the serial port is connected to an external modem. If the videoconferencing system is connected to another videoconferencing system with a computer attached, file information from the remote computer or a LAN can be sent through the digital connection that connects the two codecs.

The computers can share files and be used collaboratively, if they both have the appropriate software. The file information is displayed on the computer monitor and on other monitors in the system that are capable of showing computer output.

This design requires a computer at both ends of the conference, each running the same collaborative software. That in itself creates a problem. However, the more significant drawback is multipoint collaboration. International standard T.120 defines the means of sharing files in a multipoint mode. All of the connected codecs and the multipoint control unit (MCU) must be T.120 compatible. Since this is a relatively new standard, few networks have implemented this capability. Chapters 16 and 18 provide detailed discussions of T.120 and other standards affecting multipoint file sharing.

Another drawback to this option is a human factor, rather than a technical issue: Some sophisticated users can carry on a discussion or teach a class while operating the computer; many others cannot. In time, people will learn to use this capability, but now it is in its infancy.

12.4   Network access

Data from a LAN can be integrated into a videoconference. Most large organizations have their own data networks. If users regularly access information on the network, that capability should be made part of the visual telephony system.

A few hospitals have turned patient records into a form that can be stored electronically. If the patient's records are stored in a computer, it is helpful if both sites have access to them. If only the patient site can access the computer records, the system should allow the computer output to be transmitted to the other site.

The integration of a computer workstation into a distance education classroom provides access to computer networks. The best known network is the Internet. The computer can bring data collected on the Internet into the instruction.

If a school is equipped with a LAN, it can not only provide access to the Internet, but also allow instructors to prepare course materials at home or in their offices and simply access a server from each classroom in order to retrieve the prepared lesson material. If the remote sites are linked to the origination classroom via an enterprise network or *wide-area network* (WAN), assignments can be submitted and returned to the students over the link.

12.5 Polling systems

A few classrooms are equipped with a means of surveying the students to obtain instantaneous feedback on the effectiveness of the block of instruction. The students are asked to answer *true* or *false* or select an answer from among *A, B, C,* and *D*. They enter their responses on keypads or terminals. The responses are relayed to the instructional site through the visual telephony system or on separate links. At the instructional site, they are instantaneously analyzed and presented to the instructor on the self-view monitor. The results may be presented as raw data, statistical outcomes, or graphical representations of the data. These same systems can also be used to register student attendance, to sample attitudes toward the pace of instruction, and to signal a desire to respond. Some are data-only systems; some include data and microphone systems for student audio responses.

12.6 CD player

A significant amount of educational material as well as research documents are available on compact disks (CDs). Some hospitals are storing X-rays, microscopic images, and other types of images on CDs. Corporations sometimes use CDs to store photographs. The pictures are taken using a conventional film camera and converted into digital images for CD storage. The advantages of CDs are that they take up less space and are less susceptible to damage due to aging than is traditional paper or print storage.

If the organization uses this medium, either a CD player or a computer with a CD read-only memory (ROM) drive should be included in the system design. The computer information is integrated into the video link in the same way as computer-generated graphics (see earlier discussion in this chapter).

12.7 Video printer

A paper copy of the graphics displayed during the session can be produced with a video printer. Some printers create a photographic image of the screen. The size of the picture is 5 × 7 inches. Printers that produce larger images are available, but they are costly. Lower cost printers that produce black and white images are also available. Many users prefer these because the output can be duplicated easily and included in written reports.

12.8 Videocassette player/recorder

The output of a videotape player can be displayed locally and sent to the distant site. Prerecorded videotapes can be played as part of the course material. Distance education systems are usually designed so that the students at the distant site see the taped material rather than the instructor. At the originating site, the teacher can see both the tape and the distant students.

In some installations, the instructor is routinely recorded and archived. Students who are unable to attend a class can use the tape to review missed lessons.

Some telemedicine systems include a videotape recorder so that the consulting physician can have a record of an examination for future reference.

Videotapes made during a deposition have been allowed as evidence during a trial. Most courts require that a clock be visible in the video so that any editing is immediately apparent. VCRs with this feature are readily available.

In some cases, recording both sides of a meeting is desirable to provide a detailed record of a videoconference. Projects can be tracked. If personnel must be added to existing task groups, an historical record exists for their review. It is easy to record the combined audio from both the local site and the remote locations(s). This is done with an inexpensive audio mixer. Recording both video sources is a bit more challenging. Two separate videotapes can be created (one of the local video and one of the received video) and later edited to a single tape. Or, a technician can use an audio/video mixer to select either the incoming or outgoing video to be recorded as the discussion proceeds. Neither option is ideal.

Some videoconferencing systems incorporate a voice-activated video switcher into the system. The switcher records mixed audio and the video from the site that is speaking the loudest. The result is a record of the meeting that has the appearance of a professionally edited program.

12.9   Facsimile machine

Facsimile machines are frequently installed where visual telephony is used. Instructors can send handout materials to the remote classroom. Students can submit assignments. Patient records, as well as other pertinent information, can be sent to the consulting physician. (Of course, this material should not be allowed to sit on a facsimile machine in an area where people other than medical personnel might see it.) The physician can use a facsimile machine to send written instructions, such as diet and exercise programs, to the patient. Business meeting agendas can be modified up to the start of the meeting. Documents can be distributed for review. In the United States, documents can be signed and transmitted via facsimile. The copy is accepted as a valid signature.

Most visual telephony installations include a separate telephone line for the facsimile machine. However, it is possible to send the facsimile information as part of the visual telephony transmission. Whether the facsimile machine is connected to a separate line or to the visual telephony system, the machine should be set to answer in as few rings as possible. There is no point in delaying the conference while waiting for a document to arrive. Of course the ringer should be turned off, so that people can continue talking while the facsimile message is being transmitted.

12.10   Other equipment

The ancillary equipment discussed in this chapter is the most frequently encountered of the system additions. In telemedicine applications, specialized cameras (such as those connected to an otoscope, opthalmo-scope, or endoscope) are easily integrated into the system. An EKG can be sent over the visual telephony link and an electronic stethoscope can be built into the system.

Systems linking research facilities can include cameras mounted on microscopes or cameras that monitor processes as they take place.

The flow of products through a manufacturing process or traffic over a bridge can easily be monitored remotely and made part of a visual telephony installation.

Almost any type of human interaction or observation can take place electronically and become part of a visual telephony system.

13

Site Selection and Preparation

Fᴇᴡ ᴇxɪsᴛɪɴɢ sᴘᴀᴄᴇs are ideally suited for visual telephony. Almost all require some preparation. If you have a choice of space, which is the best place for setting up a visual telephony system?

This chapter discusses the selection of suitable space for the system, the preparation of the room, and selection of the furnishings. The principles of selecting a location for a visual telephony system and preparing the space to get the most out of the system are basically the same for all applications. The following detailed guide to setting up a videoconference room will provide a thorough analysis of the principles involved in establishing not only a system for business meetings, but systems for instruction, consultation, diagnosis, investigation, and more.

13.1 Facility size

Before searching for a location, it would be helpful to know how much space is required. This will depend on the number of people the room must accommodate and the equipment that will be used. The calculations given in the following sections provide a guideline. The layout of a videoconferencing room is not an exact science. There is plenty of opportunity for ingenuity. By following the logic used to determine the minimum amount of space required for traditional room designs, one can create an acceptable design for both ordinary and unusual spaces.

13.1.1 Width of room

The first piece of information to be considered is the number of on-camera participants. In the majority of existing videoconference rooms, the participants are seated along one side of an oval conference table. The people at the remote site are electronically on the other side of the table. The oval shape is used so that each participant is seated the same distance from the camera(s) and monitor(s) (Figure 13.1).

Table length can be determined by considering how many people will be seated at the table. The typical conference room chair is 2 feet wide. There should be a minimum of 1 foot between chairs. Allow at least an extra 6 inches at each end of the conference table. Table 13.1 reflects typical table lengths.

An aisle is necessary on at least one end of the table. This adds another 2.5 feet to the width of the room per aisle. Table 13.2 shows the minimum width of a conference room designed to seat participants along an oval table.

Alternative table shapes, such as a trapezoid (sometimes called a blunt triangle) or separate straight tables can save some space (Figures 13.2 and 13.3). Table 13.3 is for a trapezoid-shaped table. The same measurements apply to two straight tables.

13.1.2 Length of room

The participants need to be seated far enough from the camera to allow the lens to capture at least three participants in a single view. Of course, if there are seats for only one or two participants, they may sit closer to the system. The typical lens used in visual telephony has a 60-degree pick-up. Table 13.4 shows the minimum distance participants can be seated from the camera. Four participants can be

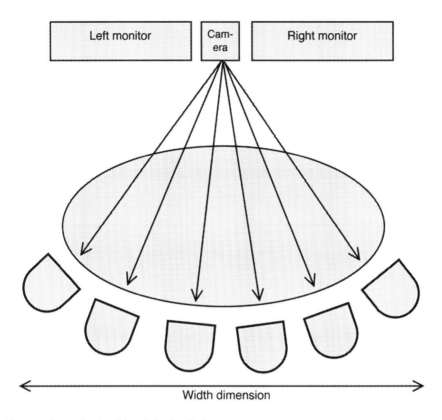

Figure 13.1 Oval table with six chairs.

Table 13.1
Table Length

Participants	Table Length (feet)
1	3
2	6
3	9
4	12
5	15
6	18

Table 13.2
Room Width for Oval Table

Participants	Table Length (feet)	Room Width With One Aisle (feet)	Room Width With Two Aisles (feet)
1	3	5.5	8.0
2	6	8.5	11.0
3	9	11.5	14.0
4	12	14.5	17.0
5	15	17.5	20.0
6	18	20.5	23.0

imaged in one view or with a split-screen configuration. Rooms designed for more than four participants must rely on either moving the camera or employing a split-screen design in order to image all of the participants.

The chairs take up 2 feet, and there must be an aisle behind the chairs. The aisle adds another 2.5 feet at a minimum (Table 13.5).

If the system equipment is located in a rollabout cabinet, add 2 feet for the depth of the equipment and 6 inches for space between the cabinet and the wall (Table 13.6).

If the conference room is a site-built design, with the equipment in a partition wall, there will need to be an equipment room behind the wall that is at least 6 feet deep (Table 13.7).

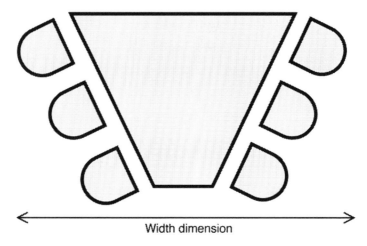

Width dimension

Figure 13.2 Trapezoidal table with six chairs.

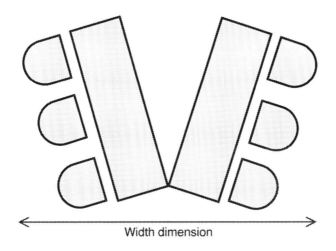

Width dimension

Figure 13.3 Two rectangular tables with six chairs.

Table 13.3

Room Width for Trapezoidal Table

Participants	Table Length (feet)	Room Width With One Aisle	Room Width With Two Aisles (feet)
2	2.5	Not possible	7.5
4	4.5	Not possible	19.0
6	7.0	Not possible	13.0

Table 13.4

Distance of Participants From Camera

Participants	Width of Camera View (feet)	Distance of Participants From Camera (feet)
1	2	2.0
2 / 4*	5	5.0
3 / 6*	8	8.0
4 / 8*	11	10.5

*First number is for a single fixed camera; second number is for a movable single camera or for a split-screen design employing two cameras.

Table 13.5
Room Length

Participants	Width of Camera View (feet)	Distance of Participants From Camera (feet)	Length of Room Including Seats and Aisle (feet)
1	2	2.0	6.5
2 / 4*	5	5.0	9.5
3 / 6*	8	8.0	13.5
4 / 8*	11	10.5	15.0

*First number is for a single fixed camera; second number is for a movable single camera or for a split-screen design employing two cameras.

Table 13.6
Room Length for Rollabout Systems

Participants	Width of Camera View (feet)	Distance of Participants From Camera (feet)	Length of Room Including Seats and Aisle (feet)	Length of Room Including Seats, Aisle, and Cabinet (feet)
1	2	2.0	6.5	9.0
2 / 4*	5	5.0	9.5	12.0
3 / 6*	8	8.0	13.5	16.0
4 / 8*	11	10.5	15.0	17.5

*First number is for a single fixed camera; second number is for a movable single camera or for a split-screen design employing two cameras.

13.1.3   Additional space requirements

Some room designs include a seating area behind the participants. Others add an observer seating area to one side of the room. Either of these options requires additional space.

In Chapter 12, we discussed the ancillary equipment typically found in videoconference rooms. If a document stand is included in the design, it should be near the participants. A small document stand can be placed on the conference table. Larger units will need a separate table. Many room designs include a table with the document stand, facsimile machine, and telephone just behind the participants. Other designs allow space at one end of the conference table.

Table 13.7
Room Length for Custom-Built Room

Participants	Width of Camera View (feet)	Distance of Participants From Camera (feet)	Room Length Including Seats and Aisle (feet)	Room Length Including Seats, Aisle, and Equipment Room (feet)
1	2	2.0	6.5	12.5
2 / 4*	5	5.0	9.5	15.5
3 / 6*	8	8.0	13.5	19.5
4 / 8*	11	10.5	15.0	21.0

*First number in each cell is for a single fixed camera; second number is for a movable single camera or for a split-screen design employing two cameras.

13.1.4 Typical conference room dimensions

The available space will seldom exactly meet the requirements calculated in the preceding tables. Taking the information just given into account, most videoconference rooms are approximately the sizes shown in Table 13.8.

Table 13.8
Typical Videoconference Room Sizes

Participants	Room Dimensions (feet)
1	Not practical*
2	Not practical*
3	12 × 16
4	15 × 12**
4	15 × 18
5	18 × 18***
6	20 × 20

*People seldom build a videoconference room for only one or two people. If space is to be allotted, the room is set up for at least three participants.
**Rooms that are too short for four participants in one camera view must use split-screen or camera movement to image the participants.
***The only rooms designed for five participants are in sites that were not large enough for a six-position room.

13.2 Access

Each room must have at least one door. Having two doors, one on either side of the conference table, is even better, enabling people to enter and leave easily during a meeting.

If there is an equipment room, access must be provided from the conference room. Another door into the equipment room from the corridor is helpful. Care should be taken to ensure that it is kept locked and that technicians cannot enter without alerting conferees of their presence. This can be done by providing an alarm light in the conference room that is illuminated when the exterior equipment door is opened.

13.3 Location

The location of the room is an important consideration. If the intended users are all top executives, it is usually a good idea to locate the room near the executive office area. If a mixture of senior level managers and other employees is expected to use the installation, it would be better to locate it in a neutral location, such as a conference center. If the room is located in executive office space, some employees will be too intimidated to schedule a videoconference.

It is always a mistake to put the videoconference room in a very obscure part of the building. It should be easy to find. The location should signify that the organization is proud of its decision to install videoconferencing. Avoid heavily traveled, noisy corridors. If a conference center location is not available, a videoconference facility can be located off a major walkway. Prominent signs should be placed along the main corridor to help people locate it and remind people to use visual telephony.

If possible, the room should be interior space. Offices with windows are prized by employees, but windows are a detriment to visual telephony. The light coming in the window will change in both color and intensity. This will cause the video to vary. The windows can be covered with drapes or blinds, but in most cases, it is better to leave the prime window space for employees and locate the videoconference room in interior space.

Another reason for avoiding space along an exterior wall is noise. Windows are poor sound barriers. Exterior traffic noise can enter the conference audio and cause a distraction.

Avoid space near elevator shafts, loading docks, heavy machinery, and air handling equipment. The ambient noise can create a serious problem with conference audio. The location should be monitored throughout the day. Some areas are noisy during some hours and quiet the rest of the day. This is particularly true of space near a cafeteria or entrance to the parking area. The walls can be treated with sound-absorbing material, but it is better to choose a quiet location and avoid this cost, if possible.

An ambient noise level of at most 45 dB [45dB SPL(A)] and preferably 40 dB should be achieved by either site selection or sound treatment.

13.4 Engineering considerations

Even the best selected space will need some extra attention to make it suitable for visual telephony. Here are some considerations.

13.4.1 Acoustic treatment

The look of the room may range from sparse to very elegant. However, even the most spartan facility must have carpeting and acoustic ceiling tiles. The room walls need to be treated with sound-absorbent material.

The front wall of the conference room, classroom, or courtroom is the wall in which the equipment is mounted (site built) or the wall behind the rollabout cabinet. This wall need not be acoustically treated. The rear wall requires attention because the microphones will face this wall and sound from the speakers will be reflected from it back to the microphones. In most installations, acoustic panels are used to deaden reverberations.

The side walls need not be covered. Since the cost of treating all of the walls is only slightly more than the cost of putting panels on just the rear wall, many organizations do cover side walls.

In some sites, drapes are used on the rear walls instead of panels. If this option is selected, the drapes should be full length, double pleated, and lined. They should also be flame retardant. The side walls may also be covered with matching material.

Room reverberation time should be between 0.3 and 0.5 second. The room's absorption coefficient should lie between 0.25 and 0.45.

13.4.2 Lighting

The majority of the room light should be concentrated on the users. Avoid hot, glaring studio-type lights. Television personalities get paid to submit to the discomfort of these; visual telephony users do not. Performers wear makeup that prevents them from looking washed out by the lights; visual telephony users do not. Also, bright lights can flare off white clothing, and there is little one can do to dictate the color of shirts and blouses users should wear to a meeting.

While users should be well lit, the monitors should not be in direct light. It is better if they are in shadow. These contrasting lighting objectives can be met by fitting the light fixtures with parabolic reflector diffusers of an angle sufficient to cast a shadow to within 30 inches of the floor along the front wall.

Fluorescent lights with a color temperature of 3,600 degrees kelvin should be used above the table. The light intensity on the table should be about 100 foot-candles or 1,000 lux. At least 60% should be on the participants.

The rear wall should be lighted and free of shadows. The rear wall will require additional lighting known as *wall wash*. The wall wash should be of half the intensity of the participant lighting and the same color temperature.

13.4.3 Air flow

The equipment used in visual telephony has a very low power consumption. All of the equipment will use less than 1,500 watts. This is roughly equal to the body heat of five people. With the potential for six or more people in the room for several hours and a number of sessions in a day, a continuous flow of fresh air must be provided. Because the number of people using the room may vary, there must be a thermostat located in the room to control the temperature.

The air exchange should be quiet, even when operating at maximum capacity. Microphones, whether located on the table or mounted on the ceiling, will be directed toward the rear wall. The outlets should be forward of the conference table, out of reach of the microphone pick-up area.

13.4.4 Power

A rollabout system should be provided with a dedicated 15-amp circuit with four 3-prong receptacles. A second 15-amp circuit should be provided for the other equipment in the room.

In some installations, a small document camera is conveniently placed on the table. There may be other pieces of equipment such as a notebook computer on the conference table. If this is the case, a safe outlet for their use should be provided.

Outlets are needed near other pieces of equipment such as a facsimile machine and VCR. These are seldom a concern since building codes require AC outlets along the walls. However, if a wall- or ceiling-mounted presenter camera is to be included in the design, a power outlet will need to be provided near it.

13.4.5 Telephone service

There should be at least one voice telephone in the room, preferably with a speaker. This can be used to quickly add someone to the conference, to call someone off-site to clarify a particular issue, or to allow a technician to work hands free with the transmission supplier and the distant end during trouble isolation. Also, there should be one analog line near the back of the equipment unit to allow for remote diagnostics. A third line may be needed for a facsimile machine.

13.4.6 Cabling

There are usually cables running from the visual telephony system to the conference table. At a minimum there are microphone cables. If a wired controller is used, there will be a cable from it to the system. There may also be cables from a document stand and a computer connection. If possible, these wires should be routed through a channel.

There may also be cables connecting other equipment to the system. For example, a presenter camera will be connected using a coaxial cable. Wherever possible, these should be run through walls or in the ceiling. If this is not possible, they must be carefully placed in the room to avoid a tripping hazard.

Suitable cables for connecting the system to the digital telephone network or network access equipment should be terminated near the system.

13.5 Decoration

Before you let your interior decorator loose in your new visual telephony installation, read these guidelines.

13.5.1 Color considerations

When the human eye is viewing a scene with multiple colors, it perceives them at different distances. Red has the shortest wavelength and therefore appears closer. Blue is the color farthest from red in the visible spectrum. It appears farther away. Because flesh tones contain a good deal of red, the use of a blue background will be consistent with the eye's viewing characteristics.

An added advantage of using a blue background is that, when the eye focuses, it does so on the edge of the object being viewed. Since blue and red are at opposite ends of the spectrum, a blue wall creates the best edges possible. The participants easily stand out from the background. The purer the shade of blue, the better. Sky blue is a very effective shade for video backgrounds.

There should be no patterns or stripes. These are visually distracting and difficult for the video codec to manage.

13.5.2 Wall decor

Decorations are also a potential distraction. However, the corporate logo is a familiar sight and can be placed on the rear wall. Many users put the geographic location of the site on the rear wall. This is particularly helpful on multipoint conferences.

13.5.3 Carpeting

The floor should be covered with a medium-density carpet. In some situations, the ceiling cannot be covered with acoustic tiles. If the ceiling is of hard plaster, more acoustic treatment is required. A dense pile carpet often adds enough absorption to compensate.

13.6 Furniture

Furniture appropriate for the task and for the electronic environment will enhance the success of the visual telephony facility.

13.6.1 Conference table

Ideally, the table should be curved or oval to allow an equal viewing perspective for all the seated participants. The distance that conferees should be seated from the camera is shown in Tables 13.3 through 13.6. This measurement is from the video equipment to the people seated along one side of the conference table. The width of the table is of no consequence. Aesthetics should dictate.

The conference table should have a matte surface so there is no glare. The optimum color is a light oak or clear pine. It should be no darker than a medium teak. Lighter colors help to reflect light upward to fill in shadows on participant faces, creating better images.

13.6.2 Other tables

There may be other tables in the room for a document stand, facsimile machine, telephone, and VCR. They should match the color and finish of the conference table.

13.6.3 Chairs

The chairs should be comfortable. People will be spending one or more hours sitting in them. They should not have highly reflective parts, such as chrome. These will cause the room lights to be reflected directly into the camera. The covering should be toward the red portion of the color spectrum and of medium density. Very dark colors will absorb light, while light colors reflect it. Sometimes empty chairs will be in view. The automatic iris of the camera is affected by the amount of light reaching it. A light-colored reflective chair will cause the iris to close and the rest of the picture to be darker than normal. A highly absorptive material will have the opposite effect. Strive for medium color densities.

13.6.4 Clock

A clock situated above the equipment is useful to keep participants aware of booking times. Often, additional clocks are added to reflect the time zones of commonly viewed distant locations.

13.6.5 Sign

A sign indicating *vacant* or *in use* should be located by the entrance door(s).

13.7 Security

The room should be secure while not in operation. Arrangements should be made for using the facility outside of normal business hours. If the room is kept locked, procedures covering access, at any time, should be available to all users, ideally as a notice displayed near the room entrance.

13.8 Applicability of design principles

The design of a visual telephony room is highly dependent on its intended use. Clearly, a videoconference room is very different from a telemedicine examining room. The guidelines contained in this chapter are useful for designing most visual telephony sites. In some cases, the chapter does not specifically address an environmental concern; such concerns will need to be reviewed with an architect. The chapter does, however, provide an appropriate checklist for a design professional to use in planning a new facility.

Part 4

Networking

CONNECTING VISUAL TELEPHONY installations is the subject of the next several chapters. We will start by understanding the architecture of the North American telephone network (Chapter 14). Next we will talk about why digital transmission is the transmission of choice (Chapter 15).

We will then digress to explore the codec, an important piece of equipment that interfaces the analog world of video with the digital network (Chapter 16). Once we have an understanding of the codec, we will examine the various options for connecting visual telephony installations through the digital network (Chapter 17).

Finally, we'll look at the intricacies of connecting more than two points in a multipoint situation (Chapter 18). Just as face-to-face communications encounters often involve more than two people, visual telephony applications often involve more than two sites.

14

North American Telephone Network Architecture

THERE WAS A time when making a telephone call involved ringing the operator and asking to be connected to another person. If it was a local call, the link was easy. Everyone in town had a direct connection from his or her home or office to the operator's switchboard. Making a call to a person in another town was more complicated.

The modern telephone network evolved from this design. Today, we can dial any of a billion telephones around the world, but the basic structure of the network connections is the same as it was in those formative years.

14.1 Local calling

Each home or business that is connected to the *public switched telephone network* (PSTN) is connected to a local switch by a pair of wires. The *switching machine* has replaced the operator. When you dial a local number, the switching machine links the two pairs or *subscriber lines* together, just as the operator did (Figure 14.1).

14.2   Beyond the local network

Just as in the early years, a call to a person connected to another switch is somewhat more complicated. Here is how it works.

14.2.1   Trunking

A call to a subscriber that is wired to a different switch than the caller's switch requires the use of a connection between the two switches. The connections between switches are called *trunks* (Figure 14.2). Trunks are permanent connections between switches and may be accomplished with wires, optical fibers, or microwave radio connections.

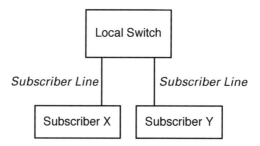

Figure 14.1   Connecting a local call.

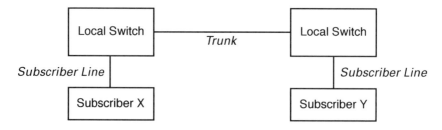

Figure 14.2   Trunks connecting local switches.

It would be impossible to connect every switch in the world with every other switch. Instead, there is a hierarchy of switches. In the U.S. telephone network, there is a five-level hierarchy.

14.2.2 Switching hierarchy

At the base of the hierarchy is the local switch we mentioned earlier. If a call is between subscribers connected to this switch, the call is completed within the switch.

If the call is to a subscriber connected to another local switch, the call is routed to a *tandem switch*. Several local switches have trunks connecting them to the same tandem switch. Calls between subscribers, each wired to different local switch, are trunked to the same tandem and handled within the tandem. Other calls are routed either to another tandem, or to a long-distance carrier (Figure 14.3).

In general, if a call is preceded by the digit *1* and an *area code,* the call will be routed to a long-distance carrier's switch. The long-distance carriers are free to structure their networks in any way they find economically feasible. However, they all use a switching hierarchy similar to the one described for the local telephone companies. Typically there are three more levels of switches. The farther apart the subscribers, the more switches through which a call routes. The highest level switch is an international gateway.

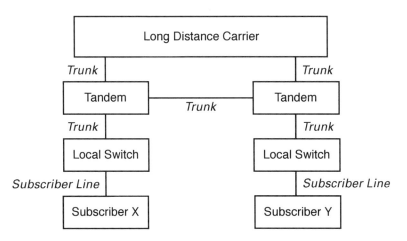

Figure 14.3 North American switching hierarchy.

For domestic calls, the long-distance carrier is supplied with the area code and telephone number of the called subscriber. It uses this information to route the call through its network to the local company tandem switch nearest the called subscriber. The tandem switch then uses the telephone number to route the call to the correct local switch. The local switch receives the called party's number and completes the connection.

On international calls originating in the United States, the caller must dial the digits *011* followed by a *country code* and *city code,* rather than a *1* and an area code. The long-distance carrier receives the country code and city code as well as the telephone number. These calls are sent to the appropriate country for completion.

14.2.3   Inter-LATA calling

In 1984, the U.S. Justice Department forced AT&T to divest itself of the local telephone companies it owned. It established a blueprint for ensuring competition among long distance carriers. The local telephone companies were left as regulated monopolies.

Under the court-ordered plan, the United States was divided into *local access transport areas* (LATAs). These LATAs were derived from the existing calling patterns. There is a large concentration of telephone traffic within the geographic boundaries of a LATA. However, LATAs do not cross state boundaries. Calls within a LATA are provided by the local telephone company or *local exchange carrier* (LEC). The local exchange carriers formerly owned by AT&T were divided into seven Regional Bell Operating Companies:

- NYNEX;

- Bell Atlantic;

- Ameritech;

- BellSouth;

- Southwestern Bell;

- U S West;

- Pacific Telesis.

The many LECs not owned by AT&T were left intact. Among the largest of these are GTE and United Telephone.

Calls between LATAs, or *inter-LATA* calls, must be handed off to a long-distance company, called an *inter-LATA carrier* or *interexchange carrier* (IXC) (Figure 14.4). Every telephone customer was required to select a *primary interexchange carrier* (PIC). Inter-LATA calls are automatically routed to the PIC. However, the customer can elect to use a different inter-LATA carrier for a specific call by preceding the call with the carrier's code. The codes for the three largest inter-LATA carriers are:

- AT&T 10 10288;

- MCI 10 10222;

- Sprint 10 10333.

This competitive model makes it possible for consumers to select the long-distance carrier of their choice and still be able to connect with any other subscriber. Table 14.1 depicts the routing between two subscribers that have selected different long-distance carriers. Subscriber X has selected Inter-LATA Carrier A for long-distance service; Subscriber Y has Carrier B.

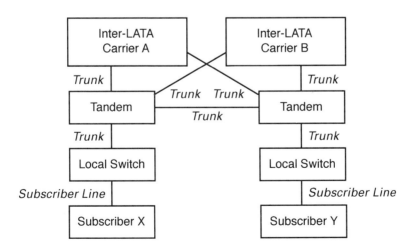

Figure 14.4   Inter-LATA calling.

Table 14.1
Anatomy of an Inter-LATA Call

1. Subscriber X picks up the telephone receiver and gets dial tone from the local switch serving Subscriber X.
2. Since Subscriber Y is in a different LATA, Subscriber X dials the digit 1 + area code + Subscriber Y's telephone number.
3. The 1 + area code alerts the local switch that this is an inter-LATA call. It is routed to a tandem switch.
4. The tandem switch routes the call to the inter-LATA carrier selected by Subscriber X.
5. The inter-LATA carrier transports the call over its network and delivers it to the tandem switch nearest Subscriber Y.
6. The tandem switch sends the call to the local switch.
7. The local switch serving Subscriber Y completes the connection to Subscriber Y.

Because the party that initiates the call is the party that is charged, Subscriber X has been able to select which long-distance carrier's service he or she wishes to use. The call is completed even though Subscriber Y has selected a different long-distance carrier. Neither subscriber is able to select an alternative LEC because at the present time there is only one for each area.

14.2.4 Bypass

The inter-LATA carriers can offer direct connections to their networks (Figure 14.5). Large customers may elect to buy a bulk offering from an inter-LATA carrier that includes a direct connection that bypasses the local switch and tandem switch. These are called *access lines*. From 1984 to 1994, the cabling was still provided by the monopoly LEC. The LEC's customer was the inter-LATA carrier. In 1994, some states opened the local cabling to competition (and in return allowed the LECs to enter into competitive businesses such as cable television). In 1995, Congress established competition in local access throughout the country.

The typical bypass connection is used by business for special telecommunications services. Figure 14.5 depicts a bypass connection between Corporation X and Inter-LATA Carrier A. Corporation X can dial Subscriber Y just as in earlier examples. However, Subscriber

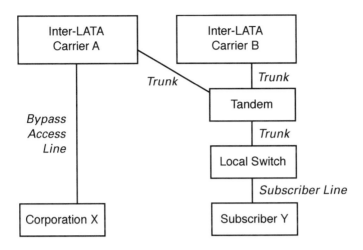

Figure 14.5 Direct access to inter-LATA carrier.

Y can only call Corporation X if the call is placed over Inter-LATA Carrier A's network.

If an organization connects to an inter-LATA carrier using a direct connection that bypasses the local switch, a distant caller cannot be connected to the line unless the entire call is on the network of the carrier that supplies the bypass. In the example of Figure 14.5, Subscriber Y could only be connected to Corporation X by dialing Inter-LATA Carrier A's access code.

This type of connection is not commonly used for traditional *plain old telephone service* (POTS). However, as we will discuss in Chapter 15, it is often encountered in visual telephony.

15

Transmission

I N THE LAST CHAPTER, we examined the architecture of the North American telephone network and discussed how the various carriers make connections across town, across the country, and internationally. Although both video and voice calls ride on this same network, the mechanics of visual telephony differ from those of voice telephony. In this chapter, we look at digital transmission, which is the basis for the transmission of computer data and visual telephony.

15.1 Analog and digital transmission

Sound is transmitted through the air as a wave of varying pressure. Light is also a *waveform,* made up of photons. Neither of these signals can be sent long distances without dissipating. Therefore, an electronic replica of the waveforms is created. This can be sent either through the air or over wires. This analogy of the original wave is known as *analog transmission.* It has been the dominant form of electronic information transfer for more than 100 years. Telephone, radio, and television are examples of analog signals.

In recent years, a new form of telephone transmission has been introduced. Rather than transmit the electronic analog of the sound and light, a sample is taken and assigned a code based on its content. The code is a series of pulses. The presence of a current is called a *one* or a *mark* and the absence of a charge is a *zero* or a *space*. (Computers use information in this form to store and process data.) The transmission facilities that carry this type of signal are known as *digital services.*

15.2   Digital structure

Chapter 14 described the trunks used to connect telephone switches. Many of these trunks are made up of two pairs of wires, one pair for sending signals in each direction (four wires in all). In the late 1950s, most of the trunks in metropolitan areas were four-wire connections.

Each trunk was used to transmit one telephone call at a time. This was not very efficient, so Bell Telephone Laboratories began work on digital transmission systems as a way to relieve the demand for underground cables used for trunks. The digital system they developed can carry 24 simultaneous voice calls over the same two pairs of wires.

15.2.1   From analog to digital

The vast majority of telephone traffic was, and still is, voice calls. Since the primary use of these trunking facilities was the transmission of voice calls, the design focused on delivering good quality in the frequency range of normal voice telephony. The frequency range for telephone circuits is from 300 to 3,500 Hz. A hertz (Hz) is a cycle per second. A kilohertz (kHz) is 1,000 cycles per second. The top end of the voice frequency range can be written as 3,500 Hz or 3.5 kHz.

To be certain of detecting a frequency in this range, the voice signal must be sampled 8,000 times per second. The sample is then converted into a seven-bit code. Telephone-quality voice transmissions therefore required the exchange of 56,000 bits, or digital pulses, per second.

8,000 samples per second × 7 bits = 56,000 bits per second

A bit per second is abbreviated bps or b/s. A thousand bits per second is kbps or kb/s. Note that the *k* for *kilo* is always in lowercase. In this book, we will show 56,000 bits per second as 56 kbps.

The transmission of a space is actually no signal. Therefore, the equipment decoding the digital information must be alerted that a seven-digit code is being sent. A *start pulse* is placed at the beginning of each code to signal the beginning. Therefore, the transmission actually consists of 64,000 bits a second. This 64-kbps equivalent circuit or *channel* has become the basic building block of digital telephony.

8,000 samples per second × (7 bits + 1 start bit) =
64,000 bits per second

64,000 bits per second = 64,000 bps = 64 kbps

As noted earlier, the objective of the Bell Labs scientists was to make more efficient use of the existing trunk cables. They found they were able to mix various calls by a method known *time-division multiplexing* (TDM) to achieve the simultaneous transmission of twenty-four 64-kbps signals.

15.2.2 Time-division multiplexing

Eight bits (one octet) from each call are sent in succession, as shown in Figure 15.1. It is necessary to keep the sending and receiving equipment synchronized. A *timing bit,* similar to the start bit, is sent before each iteration of eight bits from each of the 24 channels transmitted. The one iteration shown in Figure 15.2 is made up of 1 timing bit and 8 data bits from each of 24 channels for a total of 193 bits.

Channel 1								Channel 2								Chann...				
1	0	0	1	1	0	1	1	1	0	1	1	1	1	0	0	1	1	1	0	1

Figure 15.1 Portion of a typical T1 transmission.

Channel 0	1
Channel 1	10011011
Channel 2	10111100
Channel 3	11101001
Channel 4	10001110

Channel 22	11111101
Channel 23	11010101
Channel 24	10011010

Figure 15.2 One iteration of a typical T1 transmission.

$$
\begin{array}{ll}
1 \text{ timing bit} & = \quad 1 \text{ bit} \\
\underline{24 \text{ channels} \times 8 \text{ bits}} & = 192 \text{ bits} \\
\text{Total} & = 193 \text{ bits}
\end{array}
$$

This is repeated 8,000 times, totaling 1,544,000 bits per second:

193 bits $\times$ 8,000 samples per second = 1,544,000 bits per second

The same calculation can be made by recognizing that during each second of transmission there are 8,000 timing bits and 24 channels of 64,000 data bits.

$$
\begin{array}{ll}
8,000 \text{ timing bits} & = \quad\quad 8,000 \text{ bits} \\
\underline{24 \text{ channels} \times 64,000 \text{ data bits}} & = 1,536,000 \text{ bits} \\
\text{Total} & = 1,544,000 \text{ bits}
\end{array}
$$

This total is referred to as 1.544 *mega*, or million, bits per second and is written as Mbps or Mb/s. The *M* for *mega* is always capitalized. Thus, 1,544,000 bps = 1.544 Mbps.

15.3   Digital services designations

Digital services are available in various capacities, depending on the carrier and locations. Let's look at the common offerings.

15.3.1 T1 service

AT&T assigned letters to its transmission systems. The letter *T* was assigned to the digital facilities. Today, the term *T1* is used to indicate a 1.544-Mbps service regardless of whether it is carried over an AT&T system. The data rate of 1.544 Mbps is known as the *primary data rate* in the United States, Canada, Japan, and most of Central and South America.

15.3.2 E1 service

In Europe, the telephone administrations elected to rewire the local cable plant for digital services. They installed coaxial cable to replace the twisted pairs formerly installed for voice telephony. This allows a primary data rate of 2.048 Mbps. It is made up of thirty-one 64-kbps data channels and one 64-kbps channel for timing. The channels are numbered channels 0 (zero) through 31, with channel 0 containing the timing bits. This is the primary data rate in Europe, Australia, parts of Asia, and most of Africa. Increasingly, it is referred to as *E1*.

Although digital transmission was originally designed to relieve congestion in local cabling, it is now the most commonly used form of telephone transmission worldwide. Digital transmission may be over microwave, coaxial cable, satellite, and optical fibers, as well as copper wires.

15.4 Benefits of digital telephony

Copper wires, coaxial cables, and microwave radio can be affected by electrical interference. This causes *line noise*. In contrast to analog transmission, digital transmissions are less affected by line noise. It would be rare for extraneous electrical interference to be powerful enough to fool the digital equipment into treating a zero as a one, and a pulse is a one even if it is more power than usual.

Analog transmissions do not have a rigid format, or protocol, compared to the well-defined digital protocol. The telephone network can monitor digital transmissions more easily to detect and route around faults. Automatic reroutes take place without the caller recognizing a problem.

Another benefit is that digital transmission provides better utilization of cable plant. In other words, digital transmission reduces costs.

The higher quality, increased reliability, and cost effectiveness of digital transmission make it the transmission method of choice. All new telephone network facilities are digital.

15.5 ISDN

The original application of digital telephony was voice connections. This is still the primary use. Because computers require information in a *binary format* (two conditions: mark and space) to store and process data, digital transmissions are ideal for communication links between computers. Video can also be converted into a digital signal.

In analog telephony, carriers had to have separate transmission networks for voice, data, and video traffic. Now, digital transmissions can be carried over the same *integrated services digital network* (ISDN), the name given to the new global telephone network.

Most large corporations have all-digital telephone networks. Smaller companies are converting to ISDN services. Although most homes currently have analog telephone service, ISDN service is increasingly available to residential subscribers.

15.6 Summary

This chapter has introduced the concept of digital services. Chapter 16 describes how analog visual images are converted to digital signals for visual telephony applications. Having understood the nature of digital visual telephony, you will then be ready to read more about network options for visual telephony in Chapter 17.

16

Video Codecs

A CODER/DECODER OR *codec* is used to convert an analog signal into a digital bit stream (code) and, conversely, to format a digital bit stream into an analog signal (decode). Since video is by nature an analog medium and since digital transmission is much preferred over analog transmission, conversion is an integral part of visual telephony.

16.1 Digital video compression

Coding the signal actually increases the amount of capacity required to transmit the information. However, some codecs can lessen the amount of information that must be transmitted by a process known as *digital compression*. Digital compression makes visual telephony economically feasible.

One form of digital compression that is easy to conceptualize is based on the way analog television produces a moving image. The viewing surface of a television picture tube is coated with phosphorous. When energized by an electrical charge, the phosphorous emits light. The color is determined by the type of phosphorous and the brightness by the strength of the charge.

Electron guns focus narrow beams of electrons at the tube surface. This produces a spot of colored light. These beams are quickly moved across the screen, producing a line of visual information. The electron guns sweep the screen and return to paint the next line so quickly that our eye perceives a complete picture on the tube's viewing surface.

In the Americas and Japan, the picture, or *frame*, of video is refreshed 30 times a second to produce a seamless moving image. That means the entire picture must be transmitted 30 times a second. European television uses a 25 frames per second *refresh rate*.

Not much movement takes place from one 1/30th of a second to the next 1/30th. In fact, much of the information in a video signal is redundant. If you concentrate on a corner of the television screen during a program, you will notice that the video information remains the same for a long time. It only changes when there is a camera switch or a scene change. Consider how much duplicate video information was transmitted to keep that portion of the picture the same.

Digital codecs convert the analog video signal into a format that can be stored and processed by a computer. Therefore, one complete frame of video can be both stored in the local coder portion of a codec and also sent to the decoder of a distant codec, where it is also stored.

The coder then compares the frame of video in memory with the next frame. Only the changed information is addressed and sent to the distant decoder.

When the decoder receives the changed information, it changes its memory to reflect it. Thus, the memory in the decoder portion of a codec always contains the same picture as the memory in the distant codec's coder.

The decoder continually scans its stored memory and reformats the contents into an analog video signal that is displayed as a moving picture on a monitor or other display device.

By transmitting only the changed portion of the video, no redundant information is transmitted. Much less transmission capacity is needed. This is just one of the methods used to compress the video signal.

More sophisticated computer analyses of the video information can permit even further reductions in the amount of information that must be transmitted to produce a normal television picture. Less information sent translates to lower cost transmission services.

A codec is both a coder and decoder. Compressed video is being transmitted in both directions simultaneously. The result is interactive, motion video.

The sequence of processes the codec performs is known as an *algorithm*. Codecs must be programmed with the same algorithm to operate interactively.

16.2   International codec standards

The first commercial video codecs were introduced to the market in the early 1980s. Several manufacturers were selling equipment of their own design. Since the algorithms were developed independently, these devices would not work with one another. Thus an attempt was made to produce an international codec standard, but it was not widely supported by the American manufacturers.

At that time, a potential user was faced with incompatible algorithms, codecs designed to operate at various data rates, and networks that could not interconnect. Many organizations decided to postpone purchasing until the industry matured.

In 1984, the International Telegraph and Telephone Consultative Committee (CCITT) undertook the development of a new standard. The CCITT, which was part of the International Telecommunications Union (ITU), formally dropped the CCITT designation in 1991 and assumed the parent organization's name. Today we speak about ITU, rather than CCITT, standards.

Manufacturers from around the globe pledged to produce equipment that would work with it. The standard would allow visual telephony over current digital networks and the emerging ISDN. See Chapter 17 for a detailed discussion of ISDN services used in videoconferencing.

The basic function of a codec for visual telephony applications is digital video compression. However, a codec is more than merely a video compression and decompression device. For example, it must also code audio, and it must transmit the information over telephone lines. These functions are all defined in ITU standards. The suite of standards necessary to ensure visual telephony compatibility is contained in *Recommendation H.320.* The component standards are shown in Table 16.1.

16.3   ITU video standards

The *H.261 standard,* a component of the H.320 suite of standards, is broad enough to give manufacturers the freedom to produce equipment of varying price and performance. This standard sets out minimum performance levels that guarantee *interoperability,* and it defines higher quality alternatives. In addition, there is a provision for codecs to signal that they can operate on alternative algorithms, such as H.263 or proprietary coding. If two similarly equipped units are connected, they can switch to this mode.

The standard has provided the industry with a stable technology platform that will be viable for many years. The result has been exponential growth in videoconferencing and other applications of visual telephony.

Table 16.1
ITU H.320 Standards for Visual Telephony

Category	Recommendation	Function
Video	H.261	Video compression
	H.263	Alternative video compression
Audio	G.711	3.5-kHz audio, 48-kbps encoding
	G.722	7-kHz audio, 48-kbps encoding
	G.728	3.5-kHz audio, 16-kbps encoding
Data transfer	T.120	Application protocols
	H.223	Data link layer
Telemetry	H.281	Far-end camera control
Multipoint	H.231	Multipoint control units
	H.243	Multipoint establishment
Communication	H.221	Frame structure
	H.230	Control and indication
	H.242	Call establishment

16.3.1 Picture quality and data rate

The H.261 algorithm defines the transmission of video information at data rates from 56 kbps to 2.048 Mbps. A codec need not provide the full range of data rates to be compliant. Picture quality varies with both the data rate and the quality of the equipment used.

The overall performance of a visual telephony system is as good as the weakest component used. The codec is a major determinant of overall quality.

At *1.544 Mbps,* the best codec is capable of transmitting a picture roughly comparable to the news footage seen on broadcast television. Colleges using visual telephony for distance education frequently select equipment that will operate at 1.544 Mbps. Many telemedicine applications also require this picture quality. Few corporate applications require this high data rate.

You should note, however, that not every codec that operates at 1.544 Mbps is capable of producing high-quality images. This level of performance requires complex signal treatment and circuitry that can process a vast amount of information very quickly. Some codec producers keep manufacturing costs down by using simpler designs and components that are not capable of working to this performance level. Since the segment of the market that requires this level of performance is relatively small, this may be a sensible decision for some manufacturers.

However, rather than accept the smaller potential market implicit in this decision, a few manufacturers design their codecs to be able to transmit at a range of operating data rates, including high rates, even though the product cannot process video at the high data rates. The codecs process video at a lower rate and transmit no additional information at the higher data rate. This is merely a marketing ploy.

Buyers can easily detect this deception. Request that the codec vendor demonstrate the equipment over a range of data rates. If no improvement is seen beyond a particular data rate, that is the codec's true highest operating data rate.

At *384 kbps,* a well-designed codec can provide sharp images that move with lifelike fluidity and little blurring in the moving area. This quality is acceptable for many distance education and telemedicine applications. It is the optimal data rate for large group videoconferencing and many other applications such as judicial hearings and focus groups.

The quality differences among products designed for the 384-kbps market are fewer than at 1.544 Mbps. However, there are lesser quality codecs being marketed to this segment of the user community, too. The best way to evaluate codec performance is to do a side-by-side comparison.

At *56 kbps,* the video and audio quality are not useful for most applications of visual telephony, with the possible exception of surveillance. However, when two 56-kbps or two 64-kbps channels are combined, the audio and video quality are acceptable for small group videoconferences. This data rate is also used for desktop and computer-based systems. A number of products are available in this portion of the market. Quality varies widely.

Note also that codec performance is not consistent across the range of operating data rates. Units that are clearly superior at one data rate can be less impressive at another. This is due to coding decisions the manufacturer makes. In evaluating codec performance, the equipment should be compared at the data rate that will be used once the equipment is installed.

16.3.2 Video resolution

As noted at the beginning of this chapter, the cost of transmitting the amount of digital information in a video signal makes most uncompressed applications of visual telephony infeasible. Some reduction must be made in the information transmitted. There are a number of ways to reduce the amount of information transmitted. One way the ITU standard reduces the information content is by reducing the resolution of the image.

A television signal consists of lines of visual information displayed on the surface of a picture tube. In the United States, the video signal consists of 480 *horizontal lines* of video. In Europe there are 576 lines. Each line is a continuous stream of color and brightness information.

The signal on each line is continuous. There are no separate vertical lines. However, television can only display a finite number of side-by-side color differences. The American television standard can detect

525 distinct side-by-side lines of different color. This is often quoted as *vertical lines.*

The number of lines that can be displayed determines the sharpness of the image. This is called the *resolution.* The resolution of an American television signal is 525 vertical lines by 480 horizontal lines (525 × 480).

If the vertical and horizontal lines are thought of as a grid, the picture can be described by the number of blocks of visual information formed by the grid. These blocks of information are known as *picture elements* or *pixels.* In digital video compression they are also called *pels.* The standard American video signal is therefore 525 pixels on 480 lines.

We have frequently noted that the television standard in Europe differs from the standard in North America. The standard used in the Americas (with a few exceptions) and Japan is known as NTSC video. The video standard in most of Europe, Australia, Africa, and Asia is called PAL. For visual telephony to be as accessible as voice telephony, some means of reconciling these video standards had to be devised.

In the United States, the typical home television set is not capable of processing all of the information contained in the NTSC signal. Home receivers typically display fewer than 300 pixels on 225 lines. If this quality is acceptable to millions of viewers, it would seem reasonable that it should be adequate for visual telephony. Using this logic, the ITU established a *Common Intermediate Format* (CIF) as the H.261 visual telephony standard. CIF resolution is 352 pels on 288 lines. An uncompressed North American video signal consists of 525 × 480 or 252,000 pixels of color and brightness information. The CIF transmission is 352 × 288 or 101,376 pixels. This halves the amount of information that must be transmitted. The resolution difference is not apparent to most viewers. It is, in fact, higher than they get on their home television sets.

The ITU also approved a lower resolution standard. Even low-resolution pictures appear to be sharp when viewed on small screens. For such applications, a resolution that is one-quarter of CIF (176 × 144) was approved. This is referred to as *Quarter Common Intermediate Format* or QCIF (pronounced *Q-sif*). Most computer-based systems place the video in a small window on the screen. QCIF resolution is acceptable for this type of application.

16.3.3 Frame rate

NTSC video has a *frame rate* of 30 frames per second or 30 fps. The PAL frame rate is 25 fps. H.261 allows frame rates to vary. The coder can code every frame or can drop some frames. The user may either preset a desired frame rate or the codec may vary the rate dynamically, based on the operating data rate and the amount of change from frame to frame.

16.3.4 Codec performance

In addition to lowering the resolution to CIF, codecs must further reduce the amount of information that must be coded. The amount of reduction depends on the operating data rate and the differences between video frames. The codec designer determines which approaches to use and the sequence in which they will be implemented. There are always trade-offs. The three characteristics that are impacted by these decisions are resolution, motion-handling capability, and frame rate.

Resolution
The standard permits the transmission of even lower resolution images. If the picture changes significantly, the codec can reduce the information to be transmitted by sampling larger sections of the video frame and using the average information to define the entire section. This reduces the information that must be sent, but it results in a blurrier picture.

Motion handling capability
If the operating data rate is not enough to allow the codec to process and send all the information about the motion in a picture, the codec can code less resolution in the moving portion. Often this is not apparent to viewers because their eyes do not focus on the moving images. The picture returns to full resolution before their eyes perceive the blurriness. However, this is not always the case. Some codecs take longer to return to full resolution. Others ''break up'' when the equipment is pushed to its limit.

Frame rate

The decoder will send video to the monitor at a frame rate consistent with the local television standard. In the United States, the decoder outputs the content of its buffer 30 times a second. In Europe, the refresh rate is 25 times a second. This activity is independent of the frequency with which the codec at the remote location updates the information. The information in the decoder memory may not be updated as often as the contents are scanned. The coder can reduce the amount of information to be sent by comparing every other frame to the one in memory or every third one. The frame rate can go as low as 5 fps. This results in a jerky picture. In extreme cases, the picture resembles a silent movie.

The compromise between equipment cost, operating data rate (hence, transmission cost), and codec performance is often the most critical decision a buyer makes.

16.4 ITU audio standards

The codec codes the conference audio and transmits it over the digital link along with the video. The audio and codec control information are contained in the first channel of the connection.

If only one channel is established, the audio can only take up 16 kbps. The codec control uses 8 kbps. The rest (64 kbps − 16 kbps − 8 kbps = 40 kbps, or 56 kbps − 16 kbps − 8 kbps = 36 kbps) is available for video. The video quality at these data rates is unacceptable for most applications.

At higher data rates, the entire first channel can be given over to audio and codec control. This allows the audio as much as 56-kbps transmission capacity. As discussed in Chapter 15 on digital networks, traditional telephone-quality audio can be carried in this transmission capacity. Using a coding scheme known as *pulse code modulation* (PCM), a frequency range of 3,500 Hz (3.5 kHz) can be transmitted at 56 kbps. The ITU standard for this type of coding is labeled *G.711*. All codecs complying to H.320 *must* be able to operate using G.711.

In the same digital capacity, 56 kbps, using an alternative audio compression method known as *adaptive differential pulse code modulation* (ADPCM), a 7-kHz signal can be transmitted. This coding method is defined in recommendation *G.722*. The use of G.722 is optional. If one codec is set to G.722 and the other is set to anything else, the conference will revert to G.711 audio.

At two-channel operation, using either G.711 or G.722, one entire channel is dedicated to audio and codec control, while the same capacity is made available for the much more robust video signal.

Because 3.5-kHz audio is acceptable for telephone conversations, an alternative audio coding method is available for small group conferences, desktop video, and computer-based systems. *G.728* audio codes 3.5 kHz in 16 kbps. The codec control still requires 8 kbps. The remaining 32/40 kbps is available for video. On a dual-channel 56-kbps call, the video capacity is increased from 56 to 84 kbps. Twin 64-kbps connections increase the video from 64 to 104 kbps. In either case, the video capacity is increased by at least 50%. G.728 audio is only used on one- and two-channel calls. At operating data rates above two channels (2×56 kbps = 112 kbps, or 2×64 kbps = 128 kbps), only G.711 and G.722 can be used.

As before, if one codec is set to G.728 and the other is set to anything else, the default is G.711.

Table 16.2 summarizes the H.320 audio options.

The process of compressing the video signal takes time. If the audio is not delayed by the same amount, *lip synchronization* is lost. Most codecs are built with an *audio delay circuit* to solve this problem.

Some codecs do not dynamically change the delay time based on the operating data rate. This can cause a mismatch between the audio and video at some data rates. Users find this very distracting.

The problem becomes extreme when using desktop systems that provide QCIF resolution and low frame rates. The audio is seldom

Table 16.2
Summary of H.320 Audio Options

Standard	Audio Coding Range (kHz)	Audio Data Rate (kbps)	Available for Codec Operating Rates of . . .
G.711	3.5	56	112/128 kbps and above
G.722	7.0	56	112/128 kbps and above
G.728	3.5	16	56/64 and 112/128 kbps

synchronized with the video. Holding a conference under these circumstances can be impossible for users not familiar with digital video compression. This type of equipment should be avoided for applications that involve the general public, such as interactive point-of-sale displays.

16.5 ITU standards for still-image graphics

Most in-person meetings are enhanced by the use of visual aids. In some meetings, something as simple as an agenda may be distributed. In others, a professionally prepared computer graphics presentation may be used to help the presenter convey ideas. Clearly, if a videoconference is a replacement for an in-person meeting, this capability must be available. The same is true of other applications of visual telephony. Distance education and judicial proceedings, for example, cannot be imagined without some means of displaying graphical information.

During an in-person meeting, the participants can look at the visual aid, the presenter, or other participants. For a video meeting to be effective, these same options should be available.

16.5.1 File transfers

To transfer data files, both systems must be equipped with a PC to store files. The *ITU T.120* recommendation is a group of standards that defines how computer files can be transmitted along with motion video. Files containing visuals that will enhance the meeting can be displayed during the conference while motion video of the participants continues to be exchanged. In this way, participants have the same viewing options as they would during a face-to-face meeting.

The computer files can be text, spreadsheets, a graphics presentation, or any file the computer can display.

The speed with which the file is transferred depends on:

■ The amount of information in the file;

■ The operating data rate of the session;

■ The file transfer rate.

Most codecs allow the user to vary the file transfer rate during the transmission. At a slow rate, there is no noticeable impact on the video. As the transfer rate increases, the video can be affected. At an operating data rate of 112/128 kbps, the file transfer rate may cause the motion video to pause. Files can also be transferred prior to the conference and displayed at both sites simultaneously during the conference.

16.5.2 Screen sharing

This type of file transfer is not unique to visual telephony. Two computers equipped with modems can share data files using the *X modem* capability of the communications program supplied with the modem.

One limitation of this type of interaction is that the computers must both have the program that runs the file. For example, if a graphics presentation is created on Microsoft PowerPoint^{TM} and the file is transferred to another PC, it too must have PowerPoint to run the file. Another limitation is that during the presentation, the files may not be altered interactively.

Some programs available on the market share computer screens rather than merely transfer files. These can be used interactively. This allows both locations to view the information in the file and interact with it. However, after the interaction, only the site with the original file retains a copy. The originator is also the only person with a record of the alterations made during the conference.

For collaborative work groups, shared programs are better than mere file transfers.

16.5.3 JPEG

The ITU has defined a more global approach. The *Joint Photographic Experts Group* (JPEG) has specified a standard for the coding of still images. By adopting this standard, any file can be coded into a JPEG file and transferred to a distant codec that conforms to the standard. Using the *T.126* standard from the *T.120* group, both sites can view the file, annotate it, and retain the results of their interaction.

JPEG files can be created from both computer files and video images. The files can be loaded into a computer, where the presentation can then be converted to JPEG. The entire JPEG presentation can be retained in the computer or it can be stored on a disk until needed.

Here's an example: A presenter creates a PowerPoint presentation using a notebook computer. The presentation is retained, as a file, on a floppy disk. Before giving the presentation, the presenter loads the file into the computer in the visual telephony system, where the PowerPoint file is converted into a JPEG file. The presenter also uses the document camera to image objects. These pictures are also converted to JPEG format and integrated into the presentation. The entire JPEG presentation is retained as a JPEG file on the floppy.

Most videoconferences or instructional classes include some preliminary activities such as introductions, a review of the last meeting or class, and an agenda of what is to be accomplished. During this time, the presenter transfers the JPEG file to the computer in the remote system using a small portion of the transmission capacity. The impact on the motion video is negligible.

With the presentation now resident in both systems, participants at both sites can view the presentation, even though the computer at the remote site has only, let's say, Harvard Graphics^{TM} presentation software and not PowerPoint software.

During the session, both the presenter and attendees at both sites can alter the files by annotating them. Both sides retain a copy of the altered JPEG files. At the conclusion of the conference, the presenter reformats the JPEG files into PowerPoint and archives the presentation on a floppy. The remote site converts the JPEG files into Harvard Graphics and archives them on a LAN server.

By using the JPEG format, presentations, instruction, and supporting documentation for other applications can easily be shared among the locations involved.

16.5.4   Earlier still-image methods

In addition to the T.120 standards, codec manufacturers provide a form of still-image transfer. This involves a video frame storage capability in both the sending and receiving codec. The motion video is interrupted briefly while a still image is sent to the frame stores in both codecs. If a system has two monitors, the still image and motion video can be displayed simultaneously. At sites with only one monitor, there are two options. The user can view the still image with the motion video in a picture-in-picture window, or the user can view either image on the full screen with no picture-in-picture, toggling between the two.

Not all codecs offer still-image capability. Those that do may offer one or more resolutions. The video resolution of these still images can be QCIF (176 × 144), CIF (352 × 288), or 4×CIF (704 × 576). Verify that the codec you are considering can exchange still images at all of these resolutions if connectivity with other users is an important consideration.

16.5.5 NTSC versus SVGA

JPEG is an elegant way to integrate computer-generated text and graphics into a meeting that is held using visual telephony. The T.120 standards provide a means of transferring files during the conference and working with them interactively. However, as noted earlier, computer-to-computer connections can be made during a conference without using T.120. Even the earliest codecs had the ability to transmit data along with the audio and video information.

Although this sounds quite straightforward, there is a problem. Computers use a higher resolution monitor than NTSC video. *Visual graphics array* (VGA) resolution is 1,024 × 768, while *super VGA* (SVGA) is 1,280 × 1,024. If the computer output is to be integrated into the conference, both sites require a monitor large enough for all participants to see. Displaying the output of a computer requires a monitor that can display VGA or SVGA. At the present time, SVGA monitors are available only as large as 17 inches in diameter. These SVGA monitors are used as display devices for millions of PCs and have a price under $300. If a larger display is required, a *multiscan monitor* (NTSC, PAL, and SVGA) is required. These are more expensive than the simpler SVGA monitors.

Visual telephony system manufacturers generally incorporate a means of displaying JPEG files as NTSC. The resolution is much lower using the NTSC output, but the system price is also considerably lower. In many applications the trade-off is acceptable. Other users may elect to invest in higher priced multiscan monitors.

Another alternative that retains SVGA resolution at a lower system cost is the use of multiple, small SVGA monitors. The monitors are typically set into the conference table to avoid blocking the view of the participants. One 15-inch SVGA monitor can be placed between each two main participants.

16.5.6 Data ports

As noted earlier, all codecs have a means of sending data as well as the video and audio information. The data information flows in and out of *data ports*. The number of ports and the data transfer rates supported vary from manufacturer to manufacturer.

A small amount of the transmission capacity can be temporarily taken from the portion of the bit stream allocated to codec-to-codec control information. Using this capacity, the highest transmission rate possible for data is 9.6 kbps. Though this is a limited amount of transmission capacity, there are some practical uses for it. The most often encountered is facsimile transmissions. Documents can be exchanged during a conference without dialing a separate connection. Another use is the remote control of a device such as a VCR or auxiliary camera.

This amount of transmission capacity is insufficient for the transfer of computer files. Large amounts of data require greater transmission capacity if the information transfer is to take place in a reasonable time. For example, X-rays can be digitized and the information sent over digital lines. If this is a part of a telemedicine installation, it can be assumed that the physician is reading the X-ray during an examination or as a procedure is being performed. (Otherwise, the X-ray could be sent without a video link.) The transmission of an X-ray with film-quality resolution would take 8 hours at 9.6 kbps. Clearly this is impractical.

T.120 file transfer protocol allows the user to specify a greater portion of the transmission capacity to the transfer of data. This capacity is taken from the amount of transmission capacity allocated to video. In the preceding example, the video might be interrupted for 3 minutes while the X-ray is transmitted using most of the T1 data capacity.

16.6 ITU standard for far-end camera control

H.281 defines one practical use for the transfer of low-speed data along with the video and audio, and that is *far-end camera control*.

Most group systems allow the user to control the system's main camera using the system controller. A conference participant can pan, tilt, zoom, and focus the camera while actively taking part in the meeting. This feature has even greater value if the distant location can also control the same camera. In this way, a person observing a presentation can keep the speaker in view without interrupting.

In distance education applications, far-end camera control enables the instructor to zoom in on a remote student. In telemedicine, it allows the diagnostician to observe a wound or condition from various vantages. Surveillance and judicial installations also benefit from this feature.

H.281 actually does more than allow the far end to control the local camera. It actually enables the remote user to select which video source is transmitted. For example, the remote site can elect to switch to an auxiliary camera to observe the audience. Distant participants can also activate the presets in a system. It would be more accurate to describe this capability as *far-end image control*. This is a powerful tool in some applications.

T.120 also allows the user to control the far end's video. Unfortunately, H.281 and T.120 far-end camera control are not compatible. Some codec manufacturers support only one method. Others allow the user to select a method. Ideally, the system should support both methods and allow the user to select a method on a call-by-call basis.

16.7 Encryption

We are all familiar with spy tales in which secret messages are coded so that the information cannot be read by anyone but the intended receiver. Digital transmissions can be coded so that eavesdroppers cannot understand the content. This process is called *encryption*.

It is very difficult to monitor the digital signal when it is being carried on terrestrial telephone lines. Nevertheless, the U.S. government requires that certain sensitive electronic communications be encrypted. A few companies have taken similar steps to secure their corporate information.

During encryption, several mathematical operations are performed on the digital bit stream to render the digital signal meaningless to anyone without the ability to reverse the operations in the order in which they were performed.

Encryption may be applied to any data stream using a separate encryption unit. The device is inserted between the data transmission equipment, such as a codec and the telephone line. However, it is more user friendly to activate encryption using the codec and room controller. The ITU has not, as yet, addressed an encryption standard and the method of activation.

American codec manufacturers offer a U.S. government approved type of encryption, known as *Data Encryption Standard* (DES). However, the implementation differs among the manufacturers. There is no compatibility between products.

The ITU effort is a low priority and may never be accomplished. Digital signals are difficult to intercept and decode. Aside from single channel surveillance applications, video connections involve at least two independent channels. These channels will be in the same cable from the user's premises to the telephone company office at both ends of the conference. However, as discussed in Chapter 14, the channels will take different paths through the telephone network. Locating them and decoding the content would require a vast amount of time and money. A few governments might have the capability to do this, but no one else. Except for organizations that take the unusual precaution of scrambling voice calls placed between sites, the encryption of video is unjustified.

16.8 Other ITU standards

There are other ITU standards concerning visual telephony. You can read about them in the next chapter, which is dedicated to an in-depth discussion of the communication networks used for visual telephony, and in Chapter 18, which deals with multipoint visual telephony.

16.9 Selecting a codec

The H.320 standard allows for a variety of price and performance options. Both features and quality can vary. This could lead to some initial confusion. However, the choices narrow quickly when alternatives are prioritized.

The foremost concern should be the vendor's reputation for after-sales support. The electronics in competing codecs is likely to be similar. By and large, the units will be very reliable. However, when a problem does arise, the vendor should be in a position to respond efficiently. The majority of problems are minor. Many are due to users being unfamiliar with the system. This is particularly true in the first few months of service. The vendor should offer a *Help Desk* or equivalent.

Most manufacturers have a range of products, each tailored to meet the requirements of a portion of the market. Accurately defining the organization's current and near-future requirements reduces the product selection to a few models. Avoid selecting product based solely on price. As noted earlier, products with the same specifications can have very different performance characteristics. Low-cost products seldom perform as well as more expensive equipment. A low-cost product may be a very poor fit. The resulting user dissatisfaction could cast a pall over all future use of visual telephony in the organization. Price comparisons are best left until last. Products should be selected according to their performance characteristics and features.

An effort should be made to see the equipment in a side-by-side comparison using the proposed cameras, monitors, microphones, and speakers. Test tapes can be used to make an objective comparison. However, the quality of the tape can influence the outcome. If a tape is to be used, the source should be of outstanding quality. A laser disk player and an action movie provide good test video. The processed video should be recorded on a professional tape machine. The taping should be done under the supervision of an objective party. If possible, a professional video house should be contracted to perform the test.

Some manufacturers will demonstrate a proprietary algorithm and argue that it is much superior to the standard algorithm. If their equipment does not provide a good-quality picture using the H.320 standard algorithms, the product should be avoided.

After reducing the field to two or three nearly identical products, price quotes can be obtained.

Often, the selection of a codec is made at the same time as the selection of a system. The ability of the vendor to support the system after it is installed is a paramount concern.

17

Video Network Options

NOW THAT WE'VE looked at the telephone network and how video codecs work, it's time to get to the heart of the subject of digital transmission. This chapter will help you to understand exactly what options users have for connecting visual telephony systems.

17.1 Analog video

A television transmission contains a great deal of electronic information. It takes the equivalent communications capacity of more than 3,500 simultaneous telephone calls to carry one video signal. The cost of this much transmission capacity is very high.

Broadcast television reaches millions of homes. The cost per viewer is low; therefore, broadcast networks *can* profitably use analog video transmissions. Visual telephony involves only a handful of participants. Even distance education applications rarely involve more than 100 participants. Although analog television produces a better image, analog television is too costly for so few users. Instead, less expensive compressed digital transmission is used. There are a few exceptions where analog television is used. The more common of these exceptions are the following:

- Analog transmission costs increase with the distance the signal travels. If the distance is short, such as between sites on a campus or within a city, analog transmission can be cost effective.

- Coaxial cable can be used to carry video, voice, and data simultaneously. A local-area network installed for voice and data can do extra duty to transport one or two visual telephony sessions as well as the traditional data traffic. This can be a very effective means of increasing productivity in a large office complex.

- Analog signals can be carried by microwave radio. Private systems must operate in a portion of the radio spectrum that has a limited range and can be disrupted by rain. However, if the transmission distance is 20 miles or less, microwave offers an inexpensive means of linking sites with video.

- There are instances when an organization is able to lease spare analog capacity for a one-time special event. Broadcasters are willing to lease capacity when it is not required for scheduled programming. The cost is low, because the lessor does not fully recover all of the costs of the transmission.

These uses of analog video represent a small portion of the overall market. In most cases, it is more practical to use compressed digital video.

17.2 Digital video

Converting an analog video signal to a digital format doesn't reduce the amount of information that must be sent. The video signal still requires the equivalent capacity of 60 T1 lines. However, as we noted earlier, a digital compression codec can reduce the amount of digital capacity required for the transmission of a video signal. Therefore, digital transmissions for visual telephony applications are *compressed digital video* transmissions.

In the previous chapter, we covered digital compression codecs in detail. This chapter deals with the digital network used to transmit the signals between codecs.

Compressed video that approaches broadcast quality is possible at primary data rates of T1 (1.544 Mbps) and E1 (2.044 Mbps). Some uses of telemedicine and distance education require this quality. Very acceptable video quality is obtained with a transmission rate equal to 25% of the primary rate. Distance education, group videoconferencing, and other applications that use 25-inch displays or larger will be best served with this quality of video. Applications using small group and individual systems can use even less transmission capacity, equal to two telephone calls (112 or 128 kbps).

There are a number of possibilities for connecting your visual telephony site with other sites on your network or off your network, some permanently wired and some taking advantage of the growing amount of switched digital services.

17.3 T1 services

The transmission of data between two locations is referred to as a *point-to-point* service. A *T1* connection is a point-to-point service that transmits 1.544 Mbps of data. As noted in a previous chapter, a T1 line consists of 24 channels of 64-kbps user data and 8 kbps of synchronizing information. There is no portion of the signal allocated for dialing information. A T1 connection is permanently wired.

17.3.1 Private networks

Organizations that have a large amount of telephone traffic between offices can connect these locations with T1 links. These point-to-point connections can be used to transmit any type of digital information for voice, data, and video applications. The transmissions may be over coaxial cable, microwave, fiber optic cables, or satellite. Such private networks are very efficient. The better quality digital transmission is less costly than equivalent voice and data services acquired individually.

17.3.2 Virtual private networks

Long-distance carriers can partition a portion of their public switched network to dedicate some transmission capacity to an individual customer. They do this by programming their switches. This partitioning creates what is known as a *virtual private network*. The details of the virtual private network offerings vary from carrier to carrier. In general, they require the customer to install a bypass connection, usually a T1 access line, from the customer's premises to the carrier's office.

The major drawback to private networks and virtual private networks is their limited connectivity. Smaller offices cannot be economically linked using point-to-point digital connections. While primary rate data services may well serve large sites, they are not always economically feasible for small branch sites. A large site served by a bypass access will not be able to videoconference with its smaller branch offices or with other organizations using the bypass access. In a completely closed environment, this is impossible. Therefore, most private networks and all virtual private networks include some ability to call from on-network sites to locations off the network.

However, the reverse should not be permitted. If sites off the network are permitted access to the network, network security is compromised. The telephone hackers that cost corporations millions of dollars annually probe for access to private networks. They then call through the network to off-network sites. The costs of the off-net calls are charged to the organization that owns the network.

17.4 Switched services

Rather than have permanently wired point-to-point connections, users can take advantage of switched services networks. Getting there is half the story, and that requires an access line.

17.4.1 T1 access

T1 service may be used to connect a customer with a telephone service supplier, either an LEC or a long-distance telephone company. In Chapter 15 we referred to this type of connection as an *access line*. Each individual channel of the T1 or E1 may be configured as a switched service such as voice, circuit-switched data, or packet data. A *multiplexer*, located on the subscriber's premises, is used to separate the primary rate bit stream into individual channels that can be used for individual calls. Some office telephone switches (known as *private branch exchanges* or *PBXs*) can perform multiplexing without an external device.

Videoconferences can be established by dialing over the channels designated as *circuit switched data* channels. The channels of the T1 are set up by call type at the time the service is ordered. Any changes to the configuration must be ordered from the long-distance carrier. It takes several days to make such a change.

17.4.2 ISDN primary rate interface

The ISDN architecture provides a more sophisticated means of channel allocation. One channel of the T1 access line is designated a *dialing* or *D channel*. The carrier's switch and the PBX use this channel to allocate calls over the remaining channels, which are available to carry voice, data, or video calls. They are known as *bearer* or *B channels*. The call establishment protocol allows any call type to be placed over these channels. This type of network access is known as *primary rate interface* (PRI). In North America, it is also frequently called a *23B + D* access line.

T1 and PRI access lines are the two types of high-capacity digital access lines used in North America. PRI lines are available throughout the world. In the United States, Canada, Japan, and most of Central and South America, a PRI line is a 1.544-Mbps service made up of 23 B channels and 1 D channel (23B + D). In Europe, Australia, and parts of Asia and Africa, the PRI access line is a 2.048-Mbps service made up of 1 synchronization channel, 30 B channels, and a dialing channel (30B + D). Many organizations use these high-capacity digital access lines to reach the global ISDN. However, there is also a need for lower cost, low-capacity digital access.

17.4.3 Switched 56-kbps service

Chapter 14 explained that the origin of digital telephony was voice calls. The information in these calls could be coded into 8,000 samples of 7 bits each. This is 56 kbps of digital information. A start pulse is required to keep the coding and decoding equipment synchronized.

When AT&T began providing low-capacity digital access lines, they relied on the start pulse to maintain synchronization of their network equipment. Though the access line actually carries 64 kbps of data, the subscriber can only transmit 56 kbps. A device known as a *digital service unit* (DSU) adds and deletes the first pulse in each octet of data.

Dialing information is sent over the access line prior to the call being established, just as it is on regular dial-up voice calls. This is called *in-band signaling*. This type of access is widely known as *switched 56-kbps* or *SW56 service*.

An SW56 access line can be ordered from any of the long-distance carriers as well as from some local operating companies. One of the types of calls that can be made over a channel of a T1 line is an SW56 call. In the United States, there is a significant base of SW56 access used for videoconferencing.

17.4.4   ISDN basic rate interface

The wires used to supply SW56 service to the subscriber are the same wires used for traditional analog telephone service (POTS). Developments that followed after the introduction of SW56 service allow the LECs the ability to carry two 64-kbps B channels and a 16-kbps D channel over the same pair of wires. To distinguish a 64-kbps B channel from a 64-kbps channel containing 56 kbps of user data, the term *clear channel 64* is sometimes used to describe B channels.

This ISDN service, with two 64-kbps and one 16-kbps channel, is known as *basic rate interface* (BRI). It is also called *2B + D* service. Since dialing information is carried over the D channel, this type of signaling is called *out-of-band signaling*.

17.4.5   Rate adaption

Because a significant amount of SW56 service is deployed in the United States, there must be a means of connecting an SW56 service to a B channel (64 kbps).

The ISDN call setup protocol allows the subscriber to request a 56-kbps connection. In routing the call to a 56-kbps access line, the network converts from 64-kbps D channel signaling to 56-kbps in-band signaling.

Conversely, when a call is placed from an SW56 access line to a PRI or BRI B channel, the network will perform the conversion and signal the equipment over the D channel that a 56-kbps call is being offered. If the equipment can switch to a 56-kbps operating mode, the call can be completed.

In either case, the equipment can only put user data in 56 kbps of the B channel's 64-kbps capacity. The equipment connected to the B channel must put a pulse at the start of each octet.

The ability to switch from 64 to 56 kbps is known as *rate adaption*. All videoconferencing codecs that conform to the ITU standard are capable of rate adaption.

Table 17.1 reflects the connectivity between the various methods of accessing the global ISDN commonly used in the United States.

The U.S. telephone network has not been completely changed over to ISDN. The BRI service that is available in many locations is a digital island of 64-kbps connectivity. Calls within the local switch can be completed at 64 kbps. Calls routed to the tandem office are generally available at 64 kbps. However, many trunks between the LEC tandem offices and the inter-LATA carrier switches are 56 kbps only. The tandem switch performs the conversion. If the equipment at both ends of the connection can rate adapt, the call is completed at 56 kbps.

17.4.6 H channels

The B channels are routed over the carrier's network as independent telephone calls. For security and network dependability, there are

Table 17.1
ISDN Access Connectivity

Connectivity	SW56	BRI	SW56 Channel—T1 Access Line	PRI B Channel
SW56	Yes	Rate adaption	Yes	Rate adaption
BRI	Rate adaption	Yes	Rate adaption	Yes
SW56 Channel—T1 Access Line	Yes	Rate adaption	Yes	Rate adaption
PRI B Channel	Rate adaption	Yes	Rate adaption	Yes

usually several routes the calls can take. For example, if three calls are placed between New York and Los Angeles, one may go directly, another may route through Chicago, while the third may go through Atlanta.

This difference in network mileage can cause a difference in the time it takes for the information in the call to reach its destination. Although the *network transit time* difference is only a matter of thousandths of a second, it is enough to cause a problem if a single block of data has been partitioned into discrete portions in order to be transmitted.

In designing the ISDN, the ITU provided an architecture that allows several B channels to be combined, by the network, and treated as a single block of data. These channels will be carried together over contiguous channels of the long-distance network. Their network transit times will be the same. These are known as *H channels*. Their capacities are as shown in Table 17.2.

At the present time, only H0 service is available in the United States. The H0 channel is the ISDN network transfer rate most applicable to videoconferencing. Earlier we noted that very acceptable video quality can be obtained with a transmission rate equal to 25% of the primary rate. Twenty-five percent of the primary rate is 384 kbps. H0 channels provide one network connection for videoconferences at 384 kbps.

Table 17.3 depicts the connectivity of the various forms of access to the global ISDN that are applicable to videoconferencing. Notice

Table 17.2

H Channel Capacities

Channel	Capacity	Derivation
H0	384 kbps	25% of the primary rate.
H10	1.472 Mbps	1.472 Mbps is the capacity of a T1 access line less the 64 kbps used for the D channel. It is the remaining capacity of a single PRI line.
H11	1.536 Mbps	A D channel can be used for 24 PRI access lines. 1.536 Mbps is the channel capacity of additional PRI lines.
H12	1.920 Mbps	1.920 Mbps is the available capacity of a European PRI.

Table 17.3

Channel Connectivity

Connectivity	SW56	BRI	SW56 Channel— T1 Access Line	PRI— B Channel	PRI— HO Channel
SW56	Yes	Rate adaption	Yes	Rate adaption	No
BRI	Rate adaption	Yes	Rate adaption	Yes	No
SW56 Channel— T1 Access Line	Yes	Rate adaption	Yes	Rate adaption	No
PRI— B Channel	Rate adaption	Yes		Rate adaption	Yes	No
PRI— HO Channel	No	No	No	No	Yes

that there is a connectivity problem between H0 channels and the other forms of access.

17.4.7 Channel aggregation

Small group, desktop, and computer-based videoconferencing usually takes place over two channels, either at 112 kbps (2 × 56 kbps) or 128 kbps (2 × 64 kbps). Since the design of the global telephone network generally allows for multiple call routes for these calls, there must be a means of keeping the two channels synchronized. ITU standard *H.221* defines a method of placing sequencing information within each channel to keep them synchronized.

The data rate of 384 kbps is preferred for videoconferences between larger groups and for other applications such as distance education. There are also applications that require broadcast-like video quality at higher data rates. This can be obtained by using the primary data rate. H channels are one means of deriving these data rates without being concerned with channel synchronization. However, as the chart in Table 17.3 depicts, a codec using B channel access to the ISDN cannot connect to a codec using H channel access.

To permit connectivity between codecs that operate over two channels and codecs that are used at higher transmission rates, H.221

extends the use of sequencing information within the channel from two channels to 30 channels. This is also called channel aggregation.

Because H channels do not have channel sequencing information, multiple channels combined using H.221 channel aggregation are not compatible with the same network capacity derived using an H channel.

$$6 \times 64 \text{ kbps (B channels)} \neq 384 \text{ kbps (H0 channel)}$$

17.4.8 Bonding

The United States has lagged in deploying ISDN. H channel transmission is limited and relatively expensive. In addition, few codecs sold in the United States are capable of performing H.221 channel aggregation for more than two channels. These facts combine to make higher data rate videoconferencing difficult to implement in the United States.

One solution was devised by the manufacturers of multiplexers. Data transmission equipment, such as a video codec, can output an H0 channel to the multiplexer. The *multiplexer* divides the data into 56- or 64-kbps portions. Channel sequencing information is added. Multiple 56/64-kbps calls are established over T1, PRI, or BRI access line(s). At the other end of the connection, the sequencing information is removed and the user data are reassembled into an H0 channel. Because codecs are capable of rate adapting, the H0 channel is often replaced with a block of serial data at 6×56 kbps or 336 kbps. This frees capacity for the sequencing information.

Many people view a multiplexer as a device that *separates* the capacity of a primary rate access line into separate channels. The activity described above is one of *combining* channels into a larger block of data. This distinction is more perceived than real, but it has led to the term *inverse multiplexing* being commonly applied to this form of channel aggregation.

The most common type of inverse multiplexing is known as *bonding*. The name is derived from *b*andwidth *on d*emand. Unfortunately, bonding is not compatible with H.221 channel sequencing. Nor is it compatible with H0 channels (Table 17.4).

To establish a call using H.221 channel aggregation, each individual channel must be dialed. An H0 channel call is one network connection. It requires one telephone number.

Table 17.4

Compatibility Issues at Higher Data Rates

Connectivity	6 B Channels—H.221	6 B Channels—BONDING	HO Channel
6 B Channels—H.221	Yes	No	No
6 B Channels—BONDING	No	Yes	No
HO Channel	No	No	Yes

This poses a problem for inverse multiplexers. Because inverse multiplexing is used to put an H0 call over multiple channels, the codec is likely to have only one number in memory, or the ability to outpulse only one number when set for an H0 call.

The inverse multiplexer manufacturers have gotten around this problem by storing the inverse multiplexer's own telephone numbers in its memory. When the inverse multiplexer is called by a compatible inverse multiplexer, the calling unit requests the additional numbers that it must dial. For example, if a person in Houston calls a person in Philadelphia, the Houston caller must key one telephone number into the videoconferencing system, and request a six-channel call. The Houston inverse multiplexer places one 56/64-kbps call to the Philadelphia inverse multiplexer. When the call is established, the Houston inverse multiplexer signals the Philadelphia unit that a six-channel call has been requested. The Philadelphia inverse multiplexer provides the Houston system with the next five telephone numbers to be called.

This problem of three incompatible methods of deriving videoconference connections above 128 kbps is confusing. However, the codec manufacturers and the inverse multiplexer manufacturers have worked together to make the problem less taxing.

Many codecs on the market are designed to operate at 112/128 kbps only. This is particularly true of small group systems, desktop units, and computer-based videoconferencing systems. These systems all use H.221 channel aggregation.

Systems designed to work at higher data rates generally can also make two-channel calls using H.221 channel aggregation.

Systems that use H0 connectivity are usually routed through an inverse multiplexer. If the call is between two sites with H0 channel capability, the call can be carried over an H0 channel. However, if the call is to a site with either SW56 or B channel connectivity, bonding or a similar method of channel aggregation is used.

The one remaining incompatibility is between codecs that use H.221 channel aggregation only to derive operating data rates above 128 kbps, and codecs that use H0 and/or inverse multiplexing. At the present time, the only means of connecting such systems is by establishing the call at 128 kbps. The ITU is currently working on a means of overcoming this problem for higher data rates. It is known as *Bonding Mode 4*.

For the time being, however, there is only one way to overcome this problem of incompatibility at operating data rates above 128 kbps, and that is the use of a device known as a *multipoint control unit* (MCU). One aspect of the way an MCU works is that a separate connection is established for each site connected to the MCU. Bonding is applied to an H0 channel by the inverse multiplexer. The inverse multiplexer at the other end of the connection removes the bonding sequencing information and reconstructs the original H0 channel. Bonding information is not passed onto the MCU. The H0 channel is unchanged. MCUs are discussed in detail in Chapter 18.

If a site connects to the MCU using H.221 channel sequencing, the MCU will signal all other codecs connected to it to switch to a mode of transmission known as *6B H0*. This 6B H0 mode places H.221 channel sequencing information within an H0 block of data. If all connected codecs have this ability, the conference can take place.

17.5 Economics and availability

Most videoconference users rely on the global ISDN for videoconference connectivity. What are the economic considerations for choosing which of many services to install?

At the beginning of this chapter, we discussed the four types of digital network access. The two high-capacity forms are T1 lines and ISDN PRI lines. Lower capacity access uses SW56 and ISDN BRI lines.

The long-distance carriers offer T1, PRI, and SW56 access lines throughout the United States. Only the LECs offer BRI lines. Because there are hundreds of LECs operating in the 50 states, the availability and pricing of BRI service varies widely around the United States.

Some LECs also offer T1, PRI, and SW56 access. Price and availability vary widely.

A few LECs offer a service called *fractional T1*. A T1 access line connects the subscriber with the nearest telephone office. Rather than paying for a full T1 channel, the subscriber contracts for the use of a portion of the channel capacity. The LEC retains the remaining capacity for its own uses. This sharing of capacity extends the benefits of T1 to organizations that might not otherwise have enough traffic to warrant digital access. In general, fractional T1 service is only offered by LECs that have not fully deployed BRI service.

17.5.1 T1 versus PRI

As we have noted, a PRI line is a T1 line with one channel dedicated to dialing. The long-distance carrier charges for D channel signaling. Therefore, a T1 will be always less costly than a PRI access line. If the organization can profit from dynamically allocating the channels, then a PRI may be the right choice. However, if the site does not have a sophisticated premises switch (PBX), there is no justification for paying the high cost of PRI service. In the United States, most high-capacity access lines are T1 services.

One other distinction between T1 and PRI access is the channel capacity. Since there is no D channel on a T1 access line, the dialing information must be contained in the channels. Therefore, only SW56 connections are available using a T1 access line. PRI lines offer switched B channels. Their capacity is 64 kbps.

Whenever a site using a T1 access connects to a site using PRI access, the codec at the PRI site must rate adapt to 56-kbps operation.

17.5.2 SW56 versus BRI

The economics are very different when SW56 access lines are compared to BRI access. Because each BRI line carries two B channels, two SW56 services must be ordered to get comparable capacity. This alone makes SW56 more costly than BRI service.

There is still another economic factor: Each BRI line requires a *digital service unit* (DSU) to put the start bits into the transmission and remove them on the receive side. The carriers require a device known as a *channel service unit* (CSU) on SW56 lines. The CSU gives the carrier a remote loopback for testing. This unit is also required on T1 and PRI lines; however, most multiplexers and inverse multiplexers incorporate a CSU.

A combined DSU/CSU costs more than $500. One is required for both SW56 access lines. In contrast, a device known as a *network termination unit* (NT1) is required on a BRI line. This device costs less than $200.

In general, SW56 lines are only feasible when BRI service is not available.

17.5.3 LEC access versus IXC access

In Chapter 14, we described the architecture of the North American telephone network. Digital access provided by the LEC follows the same design (Figure 17.1).

Subscribers can easily connect between digital access lines. However, because many digital access line are provided by the long-distance carriers, they bypass the local exchange carrier's switches.

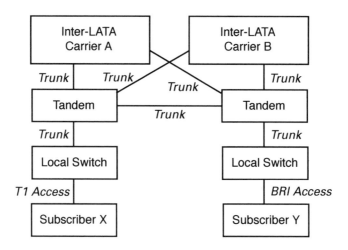

Figure 17.1 LEC-provided digital access.

As Figure 17.2 depicts, organizations connected to the ISDN using long-distance carrier-provided access lines can always call organizations that use BRI access lines. However, in order to call Subscriber X, the BRI subscriber (Subscriber Y in Figure 17.2) must have the same inter-LATA carrier (PIC) as Subscriber X uses for bypass access. Subscriber Y can also use an access code (if this information is available) to route the call to the carrier that Subscriber X uses.

The situation is much more difficult when both parties have bypass access lines provided by different inter-LATA carriers, as depicted in Figure 17.3.

Because the long-distance carriers have been in the forefront of videoconferencing, many subscribers have access lines provided by these carriers. These users can always communicate with other sites that use the same long-distance carrier. Usually that includes all of the organization's videoconferencing locations.

However, users frequently want to connect with other organizations. If the two videoconference sites are directly connected to different inter-LATA carriers, they cannot dial each other.

At the present time, the only way to overcome this problem is to use digital gateways. The three major carriers (AT&T, MCI, and Sprint) maintain gateways between each others' networks. Both subscribers must connect to the gateway. The carriers cooperate in tying their networks together at the gateway.

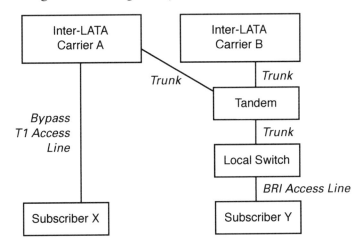

Figure 17.2 IXC bypass access and LEC access.

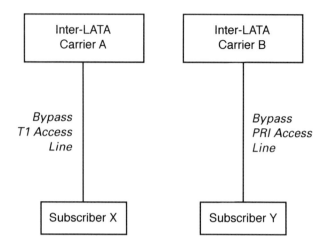

Figure 17.3   IXC bypass access for both calling parties.

This is a poor solution. Currently, all of the carriers' gateways are located in Atlanta. A party in New York wishing to connect to a site on the other side of town must pay for a call to Atlanta. The called site must also pay for a call from New York to Atlanta. In addition, there is a fee for using the gateway. The gateways must be reserved in advance. Finally, they are not always available when requested. This detracts from the user's ability to initiate a video session when it is needed. LEC-provided access is by far the best solution to ensure unrestricted interoperability.

17.6   International networks

Ironically, most of Western Europe has deployed ISDN service more fully than the United States. In Japan, there are even ISDN BRI telephone booths. Most of the other Pacific Rim countries have extensively deployed ISDN services. The United States lagged in the developed world in deploying ISDN as a result of divestiture. Recently the pace of ISDN deployment has quickened in the United States. The new competitive environment is sure to cause the LECs, which have historically moved slowly, to reassess their policies. ISDN connectivity may well be available to all customers in the near future.

Digital connections between North America and Europe and between North America and Asia are carried on undersea cables. Optical fiber cables have been laid around the world. In general, there is excellent connectivity to the developed countries. Eastern Bloc countries are slowly modernizing their telephone networks. South and Central America have ISDN service in major cities.

Those parts of the globe that are not part of the global ISDN are accessible with satellite services. This is not a solution for everyone, however, because the initial investment in Earth stations is significant.

In summary, it is theoretically possible to have video telephony anywhere in the world. The economics for some of the more remote regions, however, must be examined on a case-by-case basis.

18

Multipoint

M ANY VISUAL TELEPHONY applications require the connection of more than two sites. In such a *multipoint* session, additional considerations emerge for compatibility among sites, selecting video for display, and managing the network. This chapter takes a detailed look at the role of a multipoint control unit, or MCU, in these areas.

There are three ways of including people at several different locations in a video meeting. We'll look briefly at the first two and then see why an MCU is the best solution for most applications.

The commercial television we receive at home is an example of one method. It is a *broadcast* of video and audio programming that originates at one location and is received at many sites. There is no interaction with the receiving sites.

The second approach provides some interaction. A *multipoint audio connection* is incorporated into the program so that viewers can ask questions and make comments.

The third approach is truly interactive. *Multipoint visual telephony* allows all locations to see and be seen by all other locations. The audio is mixed so that every location can always hear the other locations.

The audio portion of a multipoint conference can be mixed without causing confusion; however, multiple video inputs cannot be mixed. There must be a way to determine which video the participants receive.

One way is to provide several video connections between sites. In a distance education installation, for example, four remote classrooms could be linked to the instructor's site by point-to-point connections. The instructor could see all four remote classrooms on four separate monitors. The students in the remote classrooms never see each other.

But if instruction is to originate from any of these remote sites, then every site must be interconnected and there must be a network control center that manages the setup and take-down of the various configurations during an instructional day. The control center adds to the cost of the network. The largest communications expense is the multiple lines that must be brought into every school. If, for example, every site can be connected to four other remote sites, four lines must be brought into all of the schools. This design is typically used when there is an optic fiber ring connecting schools in a school district or community.

Although this type of architecture might be acceptable for some distance education situations, clearly it has limited value for other applications.

18.1   Multipoint control unit

The most common method of holding multiple-location video meetings uses a device known as a *multipoint control unit* (MCU). Each site is connected to the MCU as a point-to-point connection. Technically, this is known as a *star configuration*. It may help to think of the MCU as the hub of a wheel, with each site attached by a spoke.

The MCU acts as a switcher, routing the video from one location to all of the others. Rather than viewing all the sites continually, the MCU or the participants select the site that should be seen by all the participants. The MCU routes that codec's images to all the others.

At first glance, this seems to be very simple. However, there is more to it than mere switching. A video compression codec transmits only the information that changes from one frame of video to the next. All of the codecs on the multipoint call have one frame of video in memory, which is the image transmitted by the last source location. The MCU constantly sends them information about the changes in the video coming from that site. If the MCU merely switched from one site as the video source to another, the new source location would only transmit information about the changes in its transmitted video. These changes would be applied by the other codecs to the video from the previous source location. The result would be useless.

The MCU actually determines that a switch in video source should take place. It then signals all of the codecs on the connection to prepare for it. The codecs freeze the image in their buffer, and prepare to receive a new and complete picture. The MCU makes the switch, then signals all the codecs to send a completely new picture. The complete picture from the codec selected as the source is received by all the others on the connection.

18.2 MCU protocols

The manner in which the MCU determines which location's video to transmit to all other sites is based on the protocol established at the beginning of the conference. There are four protocols, as discussed next.

18.2.1 Voice-activated switching

In this protocol, the site with the loudest audio is presumed to be the speaker's location. The video from this site is sent to all other locations. The speaker sees the previous speaker. The dialog between participants causes the MCU to switch video sources based on the relative volume from the connected locations. In other words, the person speaking the loudest is seen by everyone.

This approach does not require user intervention. In many applications, the meeting flows naturally and switching is seamless. However, this protocol can be difficult when there is a lively give-and-take of conversation among participants. Because switching is based on sound, spurious noise could cause the MCU to switch. Rather than have this happen, most MCUs are designed to measure the audio volume over a period of time, usually 1 second. The site with the highest level over the entire second is deemed to be the speaker's location. The process of switching takes another 2 seconds. Therefore, there is a delay of roughly 3 seconds from when a person begins to speak and when the participants see the speaker's image. Generally, this delay is not significant. However, when participants are freely commenting and making other sounds, the switching becomes erratic. Experienced users have found that this can be avoided by muting the microphones in the conference rooms and activating them only when a participant wishes to be heard. While this handles the problem of unwanted MCU switching, it reduces the user friendliness of the meeting situation.

18.2.2 Chairman control

For this protocol, one location is designated at the beginning of the conference to be the chairman's site. This site determines which site is seen by all others. At the start of the meeting, for example, all sites would probably see the chairman's site. The chair might then call for a report from another site and designate this site as the received site for all locations in the session. All locations except the new source would see the new source. The new source would continue seeing the chairman's site. If the meeting switches back to the chairman's location, all the other sites would see the chairman's location. The chairman's location would continue receiving video from the previous source location.

The codec at the chairman's site must have chairman control software, and the site must be equipped with a control unit. In some designs this is a keypad; in others it is a PC. The chairmanship can be passed to another location that is similarly equipped. This is done from the chairman's location.

This approach is less user friendly than voice-activated switching. It does, however, provide an experienced leader with an opportunity to hold an effective multiple-location meeting without the random glitches that often occur in a voice-activated multipoint conference.

Chairman control is the protocol preferred by most business and education users.

18.2.3 Self-selection

This control protocol is the opposite of chairman control. Here, a participant who wishes to be seen presses a button on the controller and signals this desire to the MCU. If there are multiple requests, the MCU accepts the first one. If there are two simultaneous requests, the selection is based on the order in which the sites were connected to the MCU.

The codecs at all the sites must have chairman control software and be equipped with a control unit in order to use self-selection. This approach has yet to gain popularity. However, there are some situations in which this is an ideal method.

18.2.4 Continuous view

The objective of this protocol is the continuous display of all sites. For very large conferences, that is impossible. But for most applications, with five or fewer sites participating in the conference, each location can view the other four locations in quadrants. If there are fewer than five sites, all the connected locations are always displayed in quadrants.

In some designs, a quadrant can be placed in a quadrant. In this way 16 locations can be continuously displayed. This is not recommended for most applications. It does work well, however, for surveillance.

For most applications, either locations connected after the first five receive the continuous display of the first four sites or one of the quadrants is designated to receive any site based on voice-activated switching or chairman control.

In most designs, the user can choose to see the quadrants or elect to view one of the quadrants on a full screen.

There is a variation of this approach that is ideal for distance education. Here, the instructor is at one of up to five locations. The instructor sees the four remote classrooms in quadrants while the remote sites see the instructor full screen. This approach can be further enhanced by allowing the instructor to select a remote site to be displayed, on a full screen, at all locations. For example, a student asking a question, commenting on the material, or presenting a paper would be seen by all the students. Afterwards, the instructor can then return to the instructional configuration that shows his or her image to all sites.

18.3   Cascading MCUs

In most networks, the MCU is located centrally. In some international applications, many sites from around the globe would have to call into a central MCU. This could result in multiple transoceanic connections on the same conference. Clearly, it would be far more cost effective if multiple MCUs could be linked together. This is called *cascading*.

ITU standard *H.243* defines a means for cascading MCUs. Most MCUs available today can be used in this configuration.

18.4   Dissimilar data rates

As we have noted, some applications require a data rate of 384 kbps, while other video connections can be established using only 128 kbps. An MCU switches the digital information to the codecs on the conference connection. It cannot send 384 kbps of video and audio information to a codec that is connected at 128 kbps.

In general, the MCU connection is established at a specific rate. All connected codecs must be at that rate. The conference must be established at the lowest data rate of all the systems to be connected. If the actual physical connection is at a higher rate, the MCU will signal the codecs to operate at the lowest common rate. In our example, the conference would take place at 128 kbps.

Some MCU manufacturers offer an alternative called *collaborative rates*. With collaborative rates, each site can be at a different data rate. Each port on the MCU has a codec. The video from each site is decoded and encoded separately. Thus, a site connected at 384 kbps will see all other 384-kbps sites at that data rate, but the 128-kbps site will be seen at 128 kbps and will see all the other locations at 128 kbps.

The collaborative rates alternative is useful in a number of applications. One of the most exciting is in distance education. Because many corporations now use videoconferencing, a business executive can join a distance education session from his or her desk. Leaders from government and the sciences can contribute to classes without leaving their work locations.

18.5 Audio standards

Chapter 16 discussed the three audio standards that are permitted under ITU recommendation H.320. G.711 is the only standard required by the ITU. This coding method uses one 64-kbps channel for audio and control information. The audio bandwidth is 3.5 kHz. The other more robust audio standards are optional. Therefore, when there is a disparity among the audio algorithms used by the codecs on a multipoint call, the MCU will signal all the connected codecs to switch to G.711 audio.

18.6 Interchannel synchronization

Chapter 17 discussed the three methods of deriving 384 kbps of transmission capacity. If an inverse multiplexer is connected between the telephone line(s) and the MCU, sites using bonding can call into the conference. The interchannel synchronization information is added to the codec output by the inverse multiplexer. At the receive site, the information is extracted. Therefore, the interchannel synchronization information is not passed on to the MCU. Sites that connect to the MCU using an H0 channel and sites that are connected using bonding for interchannel synchronization can be connected through the MCU (Table 18.1).

However, interchannel synchronization using ITU recommendation H.221 places the synchronization information in the codec transmission. A site that connects to the MCU over six channels synchronized using H.221 is not compatible with sites that use either bonding or H0 channels. ITU recommendation *H.243* provides a method, referred to as *6B H0,* for overcoming this problem. The MCU can signal the

Table 18.1

Interchannel Synchronization

Configuration	Synchronization
6 × 64 kbps B channels	Bonding
6 × 64 kbps B channels	ITU standard H.221
1 × 384 kbps H0 channel	None required

codecs that are using H0 channels or bonding (bonding calls reach the MCU as H0 channels) to place interchannel synchronization information into the H0 block of data. All three means of deriving 384-kbps network capacity are then compatible. This feature of an MCU can be valuable for connecting codecs on a point-to-point basis as well as on multiple-location conferences.

18.7 Data transmissions

ITU recommendation T.120 provides a protocol for transmitting data within the codec transmission path. The recommendation covers multipoint connections as well. Any site may initiate a transmission to all other sites, whether or not the site is the source location of the video transmission. The standard provides for clash control should two sites attempt to transmit simultaneously.

18.8 Audio noise

The MCU is an audio bridge. The audio from each location is mixed and sent to all other sites. In addition to the voices of participants, the audio from each site contains some noise. In poorly designed rooms, there may be a lot of spurious sounds. In rooms with good acoustics, the participants will generate some noise simply by sifting through documents and moving around in their seats. These sounds are cumulative. On a large conference, the noise can become very distracting. Some MCU designs combine the audio from the loudest four locations for retransmission to all sites. This is a very effective means of noise abatement and something that should be considered when considering the purchase of an MCU.

18.9 User MCU versus public multipoint service

Multipoint control units are designed to be located at either the user's premises or in a carrier's central office. When placed at a user's site, a digital connection from all locations on the conference must be brought to the premises. For example, a company with 12 remote offices equipped for videoconferencing may envision having no more than six on a single multipoint conference. If the operating data rate is 384 kbps, at least one-and-a-half T1 access lines must be dedicated to this application.

However, if the unit is placed in a carrier's central office, the user needs only six channels of digital access for this application. By locating the MCU at the carrier's office, there is a significant savings in access line charges. All the long-distance carriers and several local telephone companies offer multipoint conferencing using MCUs in their network.

There is a charge for the connections. Most occasional users find the fee reasonable. Frequent users find it economical to own their own MCU.

18.10 Network controller

Should an organization find it is cost effective to build a network of private T1 circuits, the MCU can serve as a visual telephony network controller. The portions of the network that are dedicated to the visual telephony application are permanently connected to one or more multipoint control unit(s). MCUs can manage several simultaneous conference connections. By requiring that all visual telephony connections be scheduled through the control center, the MCU can be used to make all the network connections. Most MCUs are configured using a PC. The PC can also be used as a connection scheduling platform. Calls can be prescheduled and established automatically. This is ideal for applications such as distance education.

Part 5

Financial Analysis

Now that you know the underlying principles of visual telephony and understand the options available to you for equipment, systems, and network services, it may be time to get on with installing a system. In some organizations, these kinds of decisions are treated as intuitive and the project is simply given a green light. For others, financial justification of some sort is required before a project is approved.

The last two chapters of this book will help you to present the financial aspects of your case for visual telephony. Chapter 19 is a discussion of the concepts that are important to a financial study. Chapter 20 walks you through some sample analyses so you can see how the theories are put to work in an actual study. Together, these chapters provide the models you need to prepare your own cost justification.

19

Cost Justification: Theory

T HIS CHAPTER EXPLAINS the financial concepts and practices used in cost justification. Once you have reviewed this information, you will be ready to proceed to the examples of cost justification provided in the next chapter.

19.1 Budgetary analysis

Most managers must operate within the constraints of a budget. A budget is both a planning tool and a financial management tool. In establishing the budget, the manager must consider what tasks are to be performed and the cost of doing them. The manager must also determine when the expenditures are likely to be made.

Visual telephony will certainly affect at least one organization's budget. It probably will affect more. The installation of a visual telephony system will create recurring costs, such as communications expenses and system maintenance costs. It may also create some savings, such as a reduction in travel costs, or some increase in revenues.

In analyzing the feasibility of implementing visual telephony, many departments may be asked to forecast the impact on their budgets. These impacts are part of the financial picture, but not the whole story. Visual telephony involves the acquisition of equipment. An asset that has an economic life beyond 1 year is called a *capital* item. A *budgetary analysis* cannot be used to analyze capital purchases. We will discuss this in more detail later. For now, consider the fact that *budgets track only annual costs,* while capital items will last several years.

Some managers get around this problem. They ask the visual telephony vendor to quote lease prices instead of purchase prices. The term of the lease is set to coincide with the economic life of the system. Lease payments are made during the year, and the benefits of visual telephony accrue during the year. If the savings or increased revenues derived from implementing visual telephony exceed the costs of visual telephony, including the lease cost, the project is viable. This is a simple and accurate way of analyzing a visual telephony proposal. However, the lease payment includes interest on the money the leasing company lays out to buy the equipment. The organization may be able to obtain better financing. Many organizations make decisions using a type of analysis that considers this and all the relevant financial factors over the project's economic life. This is a *cost–benefit analysis,* described in more detail in Section 19.2.

The other drawback of a budgetary analysis is its focus on one manager's organization. Chapter 1 discussed the widespread benefits of visual telephony. Some of the benefits identified will provide savings to a specific budgetary group. The reduction in *travel costs* is an example. Other benefits, however, are more broadly applicable. For example, one possible benefit of videoconferencing, *improved employee morale* and the resulting *reduction in personnel turnover,* avoids the cost of recruiting and training replacements. Some of this benefit will be realized by the functional manager, and some by the human resources department, which will avoid expenditures.

Other benefits contribute to increased revenues. *Faster decision making* can mean a competitive edge in the market and, therefore, higher sales volume or higher profit margins. Once again, this may affect a number of departments.

19.2 Cost–benefit analysis

A broader form of analysis is a cost–benefit analysis. Costs are easy to document. Benefits are more difficult and can be either in the form of cost avoidance or performance enhancements. Cost avoidance benefits do not affect revenues, while performance enhancements might. Because it is easier to document costs, it is easier to study the cost avoidance impact of implementing a visual telephony application rather than the performance enhancement impact. However, the revenue improvement anticipated from an application is often of more strategic importance to the organization than cost containment. When possible, include both types of benefits when performing an analysis.

19.2.1 Tangible versus intangible

Some benefits are easily quantified. Travel costs avoided can be estimated from data that are easily accessible, such as airline fares and hotel rates. Other benefits are much more difficult. What is the value to a university for improving the course offerings through distance education? In the long run, the objective is more qualified applicants, but it takes a leap of faith to attribute better applicants in the future to the implementation of distance education today. Even then, placing a value on the future applicants is nearly impossible. Yet, we must assume that the objective of *attracting more qualified applicants* is generally accepted if the project is being seriously considered.

The benefits that can be readily quantified are referred to as *tangible,* while the enhancements that cannot be assigned a value are called *intangible.* In a typical analysis, intangible benefits are referenced, but not included in the calculations. This is unfortunate because this can lead to some of the most important effects of visual telephony being ignored. This can cause projects to be delayed or even rejected.

One way of placing a value on an intangible benefit is the quantification of alternative means of obtaining the same result. Using the distance education example, the university might increase advertising, employ more admissions personnel, and make more financial aid available.

These actions can be quantified and compared to instituting a distance education program.

19.3 Financial analysis concepts

Before doing a cost–benefit analysis, some financial concepts must be understood. Some of the language used here is awkward for people unfamiliar with financial terms. The typical reader does not need to become well versed in financial analysis. In most organizations, the finance people are responsible for performing this type of study and interpreting the results. These analysts, familiar with finance concepts, will not be familiar with the application. Because they could misinterpret the data, the reader's input is important. Someone who is investigating a visual telephony project should therefore read this section once for appreciation, then use it as a reference during the preparation of an analysis.

19.3.1 Time value of money

Even in a world without inflation, a sum of money to be received in the future isn't worth as much as an equal amount currently in hand. There is, of course, the uncertainty of getting something promised at a future date. But, even if the money were guaranteed to be received, it still has less value than money currently in hand. This is because money in hand can be invested now and hence will, at some future date, be worth more than its current value. One hundred dollars placed in a bank account that pays 5.25% interest compounded annually will grow to $110.78 in 2 years (Table 19.1).

The value of a sum of money if it were in hand today is called the *present value*. The present value of a dollar in your wallet is, of course,

Table 19.1
Effect of Compound Interest

Year	Calculations
1	$100.00 × 1.0525 = $105.25
2	$105.25 × 1.0525 = $110.78

one dollar. In the example of Table 19.1, the present value of $110.78 to be received in 2 years is $100.00.

The *future value* is the worth of a sum of money to be received or disbursed at some future time. Drawing on the same example, the future value of today's $100.00, in 2 years, is $110.78.

In the example, the interest rate is 5.25%. This anticipated earning rate is referred to as the *discount rate*.

Study the relationships shown in Table 19.2, where PV = present value, r = discount rate, n = number of years, and FV = future value. Using the first formula shown in Table 19.2, we can calculate the future value, in 2 years, of $100, as demonstrated in Table 19.3.

Using the second formula of Table 19.2, we can determine the present value of $110.78 available in 2 years, as demonstrated in Table 19.4. A discount table provides an even easier way to quickly obtain the $(1+r)^n$ values. See Appendix A for a discount table.

19.3.2 Cash flow

The actual money entering or leaving an organization is known as *cash flow*. Positive cash flows are money entering the entity, and negative cash flows are money leaving.

Table 19.2
Time Value of Money

Concept	Formula
Future value	FV = PV (1+r)n
Present value	PV = FV [1/(1+r)n]

Table 19.3
Future Value of $100

Calculation
FV = PV(1+ r)^n
FV = $100(1+ 0.0525)^2
FV = $100(1.1078)
FV = $110.78

Table 19.4
Present Value of $110.78

Calculation
$PV = FV[1/(1 + r)^n]$
$PV = \$110.78[1/(1 + 0.0525)^2]$
$PV = \$110.78[1/1.078]$
$PV = \$110.78[0.9027]$
$PV = \$100$

19.3.3 Net present value

The present value of the algebraic sum of all cash flows is called the *net present value* (NPV). The present value of future cash flows is calculated using the discount rate.

Using an example similar to the previous one, let us assume an investment pays us a dividend of $100 a year. The net present value is the sum of the present values for each year. We can calculate the NPV of the next 5 years using the formula $PV = FV[1/(1 + r)^n]$ and a discount rate of 5.25%, as shown in Table 19.5.

If we were asked to determine the wisdom of investing $400 to receive the cash flows shown in Table 19.5, our analysis would look like the one shown in Table 19.6, where numbers in parentheses are negative numbers. Because the NPV is positive, we would recommend the investment.

Notice that the investment is shown as being spent in Year 0. By convention, capital purchases are treated as though they occur on January 1 of Year 1. All other expenses are shown as occurring on December 31 of each year. The use of Year 0 actually shows the capital on the books at the *beginning of Year 1*.

Altering the preceding example slightly, if we were asked to determine if an acquisition that cost $450 and returns the same cash flows is sound, our analysis would proceed as shown in Table 19.7. Because the NPV is negative, we would advise against this investment.

19.3.4 Taxes

Nonprofit organizations can use an analysis similar to the one just discussed. For many others, there is a slight complication: *The return on the investment is taxable.*

Table 19.5
NPV of $100 Dividend Annually for Five Years

Year	Calculations	Result
1	PV = $100[1/(1+.0525)^1] PV = $100[1/1.0525] PV = $100[0.9501]	PV = $ 95.01
2	PV = $100[1/(1+.0525)^2] PV = $100[1/1.1078] PV = $100[0.9027]	PV = $ 90.27
3	PV = $100[1/(1+.0525)^3] PV = $100[1/1.1659] PV = $100[0.8577]	PV = $ 85.77
4	PV = $100[1/(1+.0525)^4] PV = $100[1/1.2271] PV = $100[0.8149]	PV = $ 81.49
5	PV = $100[1/(1+.0525)^5] PV = $100[1/1.2915] PV = $100[0.7743]	PV = $ 77.43
Net Present Value		NPV = $429.97

Table 19.6
NPV Analysis for a $400 Investment, $100 Annual Return

Year	Result
0	PV = ($400.00)
1	PV = $ 95.01
2	PV = $ 90.27
3	PV = $ 85.77
4	PV = $ 81.49
5	PV = $ 77.43
Total	NPV = $ 29.97

Rather than analyze an investment, let us calculate the NPV of winning a lottery that pays $100 a year for 5 years. The winnings are taxable. Let us assume a 35% tax bracket. The formula for each year now becomes $PV = [FV - (t \times FV)] [1/(1 + r)^n]$, and is calculated in Table 19.8.

Table 19.7

NPV Analysis for a $450 Investment, $100 Annual Return

Year	Result
0	PV = ($450.00)
1	PV = $ 95.01
2	PV = $ 90.27
3	PV = $ 85.77
4	PV = $ 81.49
5	PV = $ 77.43
Total	NPV = ($ 20.03)

Table 19.8

NPV Analysis for $100 Income Annually for Five Years

Year	Calculations	Result
1	$PV = [\$100 - (0.35*100)][1/(1+.0525)^1]$ $PV = [\$100 - \$35][1/1.0525]$ $PV = [\$65][0.9501]$	PV = $ 61.76
2	$PV = [\$100 - (0.35*100)][1/(1+.0525)^2]$ $PV = [\$100 - \$35][1/1.1078]$ $PV = [\$65][0.9027]$	PV = $ 58.68
3	$PV = [\$100 - (0.35*100)][1/(1+.0525)^3]$ $PV = [\$100 - 35][1/1.1659]$ $PV = [\$65][0.8577]$	PV = $ 55.75
4	$PV = \$100 - (0.35*100)][1/(1+.0525)^4]$ $PV = [\$100 - \$35][1/1.2271]$ $PV = [\$65][0.8149]$	PV = $ 52.97
5	$PV = [\$100 - (0.35*100)][1/(1+.0525)^5]$ $PV = [\$100 - \$35][1/1.2915]$ $PV = [\$65][0.7743]$	PV = $ 50.33
Total		NPV = $279.49

19.3.5   Tax depreciation

In the previous example, we did not have an investment to consider. The purchase of an asset adds a little more complexity to the analysis.

A portion of an asset's cost is deducted as an expense each year the asset is expected to be in service. The annual expense is known as *depreciation*. Corporations usually expense an equal portion of the asset's cost each year on their annual financial statements. No cash flow results from recognizing depreciation on annual reports. However, on tax returns, businesses deduct depreciation. This lowers the amount of taxes to be paid. The amount of taxes avoided is a cash flow.

Having learned about the time value of money, we can see how unfair it is for a company to pay for an asset today and have to wait several years to get the tax reductions associated with the purchase. In recognition of this, the government allows companies to take larger depreciation write-offs in the years immediately following a purchase and lower depreciation in later years. This is known as *accelerated depreciation*.

The Internal Revenue Service (IRS) has prescribed an accelerated depreciation method known as the *Modified Asset Cost Recovery System* (MACRS). Visual telephony equipment is considered a 5-year property. The IRS schedule calls for depreciation to extend over 6 years. The portion of the asset's price that may be written off in each year is shown in Table 19.9. This table is also reproduced as Appendix B, Depreciation Factors for Tax Purposes.

If a company buys an asset that costs $400, the annual depreciation expense would be calculated as shown in Table 19.10.

Table 19.11 shows how to calculate the cash flow (CF) from the depreciation write-off, assuming a 35% tax bracket. The formula is $CF = depreciation \times tax\ rate$. This total cash flow is equal to taxes avoided because of the depreciation.

Table 19.9
IRS MACRS Depreciation Rates

Year	Depreciation Allowed
1	0.2000 or 20.00%
2	0.3200 or 32.00%
3	0.1920 or 19.20%
4	0.1152 or 11.52%
5	0.1152 or 11.52%
6	0.0576 or 05.76%
Total	1.0000 or 100.00%

Table 19.10

Depreciation Calculations for a $400 Asset

Year	Calculation	Depreciation
1	(0.2000)($400)	$ 80.00
2	(0.3200)($400)	$128.00
3	(0.1920)($400)	$ 76.80
4	(0.1152)($400)	$ 46.08
5	(0.1152)($400)	$ 46.08
6	(0.0576)($400)	$ 23.04
Total		$400.00

Table 19.11

Cash Flow From Depreciation of $400 Asset

Year	Calculation	Cash Flow From Depreciation
1	CF = $ 80.00(0.35)	$ 28.00
2	CF = $128.00(0.35)	$ 44.80
3	CF = $ 76.80(0.35)	$ 26.88
4	CF = $ 46.08(0.35)	$ 16.13
5	CF = $ 46.08(0.35)	$ 16.13
6	CF = $ 23.04(0.35)	$ 8.06
Total		$140.00

19.3.6   Cost of capital

In evaluating projects, most companies use a *weighted average cost of money* to discount net cash flows. This is derived by taking the portion of the capital derived from debt (usually 25% or less) and adjusting its nominal interest rate for taxes, then combining this with the desired return on equity (usually about 15%). The weighted average formula, where r = interest rate, t = tax rate, and E = return on equity, is

$$
\begin{aligned}
\text{Debt} &= 0.25(r)(1 - t) \\
+\,\text{Equity} &= 0.75(E) \\
\hline
\text{Weighted Average} &= [0.25(r)(1 - t)] + [0.75(E)]
\end{aligned}
$$

If we assume an interest rate of 6%, a tax rate of 35%, and a return on equity of 15%, the cost of money allocated to a project would be calculated at 12% (Table 19.12).

19.3.7 Other discount rates

Government agencies often use a discount rate that reflects the alternative investment opportunities of their constituents. In other words, if the taxpayer can earn 10% in secure investments, then the state should be sure to get at least that amount. Nonprofit organizations use a similar earnings target.

19.4 Sample net present value analysis

We now have all the information necessary to perform a net present value analysis of a business decision. Let us determine the feasibility of investing $400 in a machine that will produce $100 in additional revenue each year for 5 years. Once again, we will assume a tax bracket of 35% and a weighted average cost of capital of 12%. The machine will have the same tax depreciation as visual telephony equipment.

The cash flow in Year 0 is simply the outflow of $400 (Table 19.13).

In Year 1, there is additional revenue of $100. This is taxed at the rate of 35%. The additional after-tax revenue is $[\$100 (1 - t)] = \$ 65$ (Table 19.14).

There is also a depreciation write-off that results in lower taxes. The amount of depreciation is $(\$400)(0.2000) = \$80.00$ This results in lower taxes, in the amount of $(\$80)(0.35) = \$28.00$, which is

Table 19.12

Weighted Average Cost of Money

Item	Calculation	Result
Debt	(0.25)(0.06)(0.65)	0.01
Equity	(0.75)(0.15)	0.11
Weighted average		0.12

Table 19.13
Net Cash Flow Analysis—Step 1:
For a $400 Investment Earning $100 Annual Revenue

Year	Entry	Cash Flow	Net Cash Flow
0	Capital purchase	($400.00)	($400.00)

Table 19.14
Net Cash Flow Analysis—Step 2:
For a $400 Investment Earning $100 Annual Revenue

Year	Entry	Cash Flow	Net Cash Flow
1	Additional revenue after taxes	$65.00	

another positive cash flow (Table 19.15). The net cash flow is the algebraic sum of the two, or $93, as calculated in Table 19.16.

Table 19.17 shows how the same calculations can be done for **Years** 2 through 6.

19.4.1   Calculation of net present value

From the net cash flows, we can calculate the NPV, as is done in Table 19.18.

The *cumulative NPV* or the NPV of this investment is negative at the end of the study period. The NPV of the investment is ($64.96). In other words, the firm's stockholders will be better off if the company *does not* invest in this capital expenditure.

Table 19.15
Net Cash Flow Analysis—Step 3:
For a $400 Investment Earning $100 Annual Revenue

Year	Entry	Cash Flow	Net Cash Flow
1	Additional revenue after taxes	$65.00	
	Reduced taxes from depreciation write-off	$28.00	

Table 19.16

Net Cash Flow Analysis—Step 4:
For a $400 Investment Earning $100 Annual Revenue

Year	Entry	Cash Flow	Net Cash Flow
0	Capital purchase	($400.00)	($400.00)
1	Additional revenue after taxes	$ 65.00	
	Reduced taxes from depreciation write-off	$ 28.00	$ 93.00

Table 19.17

Net Cash Flow Analysis—Step 5:
For a $400 Investment Earning $100 Annual Revenue

Year	Entry	Cash Flow	Net Cash Flow
0	Capital purchase	($400.00)	($400.00)
1	Additional revenue after taxes	$ 65.00	
	Reduced taxes from depreciation write-off	$ 28.00	$ 93.00
2	Additional revenue after taxes	$ 65.00	
	Reduced taxes from depreciation write-off	$ 44.80	$109.80
3	Additional revenue after taxes	$ 65.00	
	Reduced taxes from depreciation write-off	$ 26.88	$ 91.88
4	Additional revenue after taxes	$ 65.00	
	Reduced taxes from depreciation write-off	$ 16.13	$ 81.13
5	Additional revenue after taxes	$ 65.00	
	Reduced taxes from depreciation write-off	$ 16.13	$ 81.13
6	Reduced taxes from depreciation write-off	$ 8.06	$ 8.06

We can redo this analysis and change the revenue assumption such
that the investment will result in an annual increase in revenue of
$150. The net cash flows would be as shown in Table 19.19. The
NPV calculation is shown in Table 19.20. You can see that this time
the project has a positive NPV. Therefore, the firm's stockholders will
be better off if the company invests in the capital acquisition.

Table 19.18

NPV Analysis for a $400 Investment
Earning $100 Annual Revenue

Year	Net Cash Flow	Discount Factor @ 12%	Present Value	Cumulative Net Present Value
0	($400.00)	1.0000	($400.00)	($400.00)
1	$ 93.00	0.8929	$ 83.04	($316.96)
2	$109.80	0.7972	$ 87.53	($229.43)
3	$ 91.88	0.7118	$ 65.40	($164.03)
4	$ 81.13	0.6355	$ 51.56	($112.47)
5	$ 81.13	0.5674	$ 46.03	($ 66.44)
6	$ 8.06	0.5252	$ 4.24	($ 62.20)

Table 19.19

Net Cash Flow Analysis for a $400 Investment
Earning $150 Annual Revenue

Year	Entry	Cash Flow	Net Cash Flow
0	Capital purchase	($400.00)	($400.00)
1	Additional revenue after taxes	$ 97.50	
	Reduced taxes from depreciation write-off	$ 28.00	$125.50
2	Additional revenue after taxes	$ 97.50	
	Reduced taxes from depreciation write-off	$ 44.80	$142.30
3	Additional revenue after taxes	$ 97.50	
	Reduced taxes from depreciation write-off	$ 26.88	$124.38
4	Additional revenue after taxes	$ 97.50	
	Reduced taxes from depreciation write-off	$ 16.13	$113.63
5	Additional revenue after taxes	$ 97.50	
	Reduced taxes from depreciation write-off	$ 16.13	$113.63
6	Reduced taxes from depreciation write-off	$ 8.06	$ 8.06

Table 19.20
NPV Analysis for a $400 Investment
Earning $150 Annual Revenue

Year	Net Cash Flow	Discount Factor @ 12%	Present Value	Cumulative Net Present Value
0	($400.00)	1.0000	($400.00)	($400.00)
1	$125.50	0.8929	$112.06	($287.94)
2	$142.30	0.7972	$113.44	($174.50)
3	$124.38	0.7118	$ 88.53	($ 85.97)
4	$113.63	0.6355	$ 72.21	($ 13.76)
5	$113.63	0.5674	$ 64.47	$ 50.72
6	$ 8.06	0.5252	$ 4.24	$ 54.95

19.4.2 Payback

In the example just given, the cumulative NPV turned positive and remained positive in the fifth year of the analysis. This is called the *discounted payback period*. The 5-year discounted payback means the stockholders will not be rewarded for making this investment until the end of its economic life. Since there is much uncertainty in financial projections, many organizations require a discounted payback of 3 years or less. This project would not be funded in such a company.

The preceding calculations are easily performed with a calculator. Decision support software such as Lotus 1-2-3 and Excel make analyses like this possible without time-consuming manual calculations. However, not long ago few companies had the resources to devote to this type of analysis. Major undertakings might have been studied with this detail, but most business decisions were not. Instead, the expected cash flows were analyzed to determine when the project returned its initial investment. In Tables 19.21 and 19.22 we look again at the net cash flows we calculated in the preceding two examples.

We compare the two assumptions in Table 19.23. Notice that this analysis gives a positive payback for both projects. However, organizations that use this approach would have required a payback of no more than 3 years to compensate for the fact that this method is not as stringent as more exact forms of analysis. Using a simple payback, neither project would be funded (note negative values at end of third

Table 19.21

Net Cash Flow Analysis for a $400 Capital Purchase,
$100 New Revenue per Year

Year	Entry	Cash Flow	Net Cash Flow
0	Capital purchase	($400.00)	($400.00)
1	Additional revenue after taxes	$ 65.00	
	Reduced taxes from depreciation write-off	$ 28.00	$ 93.00
2	Additional revenue after taxes	$ 65.00	
	Reduced taxes from depreciation write-off	$ 44.80	$109.80
3	Additional revenue after taxes	$ 65.00	
	Reduced taxes from depreciation write-off	$ 26.88	$ 91.88
4	Additional revenue after taxes	$ 65.00	
	Reduced taxes from depreciation write-off	$ 16.13	$ 81.13
5	Additional revenue after taxes	$ 65.00	
	Reduced taxes from depreciation write-off	$ 16.13	$ 81.13
6	Reduced taxes from depreciation write-off	$ 8.06	$ 8.06

year in both examples). Clearly, the investment that returns $150 a year is marginal.

19.4.3   Internal rate of return

The two examples just studied are for 5-year projects. One is clearly not advisable. The other is marginal. If a manager had only two projects to select from, comparing these two is very easy. But, in the real world, there are many proposed projects competing for a limited amount of capital. The investment alternatives will have different economic lives and different cash flow patterns. For example, a plant expansion may require outlays of cash in several years. NPV analyses of these different alternatives are difficult.

One means of selecting from among numerous projects is the use of *rate of return analyses*. The proposed capital purchase is viewed as an investment. The derived cash flows are treated as interest or

Table 19.22

Net Cash Flow Analysis for a $400 Capital Purchase,
$150 New Revenue per Year

Year	Entry	Cash Flow	Net Cash Flow
0	Capital purchase	($400.00)	($400.00)
1	Additional revenue after taxes	$ 97.50	
	Reduced taxes from depreciation write-off	$ 28.00	$125.50
2	Additional revenue after taxes	$ 97.50	
	Reduced taxes from depreciation write-off	$ 44.80	$142.30
3	Additional revenue after taxes	$ 97.50	
	Reduced taxes from depreciation write-off	$ 26.88	$124.38
4	Additional revenue after taxes	$ 97.50	
	Reduced taxes from depreciation write-off	$ 16.13	$113.63
5	Additional revenue after taxes	$ 97.50	
	Reduced taxes from depreciation write-off	$ 16.13	$113.63
6	Reduced taxes from depreciation write-off	$ 8.06	$ 8.06

Table 19.23

Net Cash Flow Comparison Analysis for $400 Capital Purchase

	$100 Revenue per Year		$150 Revenue per Year	
Year	Net Cash Flow	Cumulative Net Cash Flow	Net Cash Flow	Cumulative Net Cash Flow
0	($400.00)	($400.00)	($400.00)	($400.00)
1	$ 93.00	($307.00)	$125.50	($274.50)
2	$109.80	($197.20)	$142.30	($132.20)
3	$ 91.88	($105.32)	$124.38	($ 7.82)
4	$ 81.13	($ 24.19)	$113.63	$105.81
5	$ 81.13	$ 56.94	$113.63	$219.44
6	$ 8.06	$ 65.00	$ 8.06	$227.50

dividends, and the interest rate needed to generate these cash flows is calculated. This rate can be thought of as the investment's earning rate. It is called the project's *internal rate of return* (IRR).

Projects earning less than the cost of capital are rejected. For this reason, in this type of analysis, the cost of capital is sometimes called the *hurdle rate.* Acceptable projects are then rank ordered by the IRR. An undertaking will be approved only if capital is still available after all projects with higher returns have been funded.

Financial calculators do IRR calculations. Decision support software also calculates the IRR of analyses. We will use a graph to depict the determination to help explain the concept. Note, however, that graphing the IRR is time consuming and not recommended when performing an actual analysis. Rely on a financial calculator or software program instead.

If we take the net cash flows of the $150-per-year project (the second assumption in the previous example), we can determine the NPV using a number of discount rates (Table 19.24).

If we plot the NPV figures on a graph, as is done in Figure 19.1, we see that the discount rate required to achieve a zero (or breakeven) NPV is around 17%. Using a financial calculator, we could much more quickly determine the IRR to be 17.15%.

Table 19.24

NPV Calculations at Various Discount Rates for a $400 Capital Purchase, $150 New Annual Revenue

Year	Net Cash Flow	Discount Rate = 11%	Discount Rate = 14%	Discount Rate = 17%	Discount Rate = 20%
0	($400.00)	($400.00)	($400.00)	($400.00)	($400.00)
1	$125.50	$113.06	$110.09	$107.26	$104.58
2	$142.30	$115.49	$109.50	$103.95	$ 98.81
3	$124.38	$ 90.95	$ 83.96	$ 77.66	$ 71.98
4	$113.63	$ 74.85	$ 67.28	$ 60.64	$ 54.80
5	$113.63	$ 67.44	$ 59.02	$ 51.83	$ 45.67
NPV		$ 61.79	$ 29.84	$ 1.35	($ 24.16)

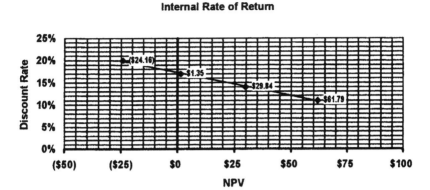

Figure 19.1 Plotting the internal rate of return.

19.5 Models

Armed with the information in this chapter, we can now evaluate alternatives for a variety of situations where visual telephony might be considered. The next chapter shows some sample analyses that will help you to construct your own analyses.

20

Financial Analysis of Visual Telephony

FINANCIAL ANALYSES OF the costs and benefits of proposed visual telephony projects range from very simple exercises to more complex studies, depending on the type of organization involved. The simplest study is one that does not include tax considerations. Government agencies, schools, and other nonprofit organizations use this type of analysis. In this chapter, we work through a simple financial analysis for a proposed distance education project. We also complete a more complicated analysis for a corporate videoconferencing project, which involves taxes, depreciation, internal rates of return, and other considerations.

20.1 Accounting for costs

The benefits derived from instituting a visual telephony network vary with the type of application, the organization, and the environment.

The treatment of costs, however, is similar for all applications. We can begin by examining typical costs and observing how they are presented in a financial analysis. Here are typical cost items for a visual telephony project:

- *Equipment cost.* This is the cost of the system. It is a capital expenditure and is therefore reflected as an initial outflow of cash.

- *Installation and shipping.* The installation and shipping costs are usually capitalized. However, some organizations treat them as operating expenditures. In either case, they are cash outflows that take place at the start of the analysis.

- *Furniture.* Room furnishings, such as conference tables and chairs, are capital expenditures.

- *Room modifications.* Simple room improvements, such as painting or window covering, can be expensed. More extensive modifications, such as carpeting and rerouting of air ducts, are capitalized. In an extensive remodeling that includes both routine maintenance and property improvements, the entire cost is capitalized.

- *Maintenance.* Even the best equipment malfunctions from time to time. Most companies buy an extended warranty on the equipment. The typical cost is 10% of the original equipment cost.

- *System support costs.* A few people will need to be trained to operate the system and help users become familiar with the system. A "rule of thumb" for estimating the annual salary for system support is to assume 20% of the system's purchase price. In addition to salary costs, there are other costs from having people on the payroll, such as Social Security tax, unemployment insurance, and workmen's compensation insurance. These additional costs add 50% to the nominal salary. This raises the general rule of thumb amount to 30% of the system's cost.

- *Network access.* Certain applications use dedicated lines to connect systems directly, in which case the dedicated line cost is reflected as an annual expense. Most applications use the global ISDN and require only a digital access line between the systems and the network. This results in an access line expense and a usage expense (see next paragraph). If digital access lines, such as BRI lines, are used *exclusively* for the application, their *full* costs should be recognized as annual expenses. Some digital access lines can be shared by applications. For example, an ISDN PRI line may be terminated in an on-premises telephone switch, known as a PBX. The digital capacity can be shared among voice, data, and video applications. If the access line is *shared* among several applications, a *portion* of the total cost of the access line should be charged to the visual telephony application.

- *Usage.* In addition to the cost of access lines, there is also the cost for connecting systems over the ISDN. The long-distance carrier can provide the costs for the anticipated use of the digital network.

Later we will discuss benefits. Be sure to use the same usage assumptions when calculating both costs and benefits.

20.2 A simple study of distance education

We can now prepare a cost analysis for a proposed university distance education system. Because universities are tax exempt, the study is relatively simple.

20.2.1 Cost worksheet for implementing distance education

Here are some cost assumptions and a step-by-step look at how to create a typical distance education cost worksheet.

- *Equipment cost.* The initial installation will be five classrooms, all combination origination and receive sites, at a cost per classroom of $75,000. There will also be an MCU to combine the sites for multiple-location instruction. The MCU will cost $100,000.

- *Installation and shipping.* These costs of $16,000 and $6,000, respectively, will be expensed.

We can begin the worksheet by inserting these known costs in Years 0 and 1 of the worksheet shown in Table 20.1. Notice that the capital expenditure is shown in Year 0, while the expense is shown in Year 1. By convention, capital expenditures are treated as though they occur a year earlier than expenses that take place the same year.

There are some other items to be purchased:

- *Furniture.* Each classroom will need $5,000 worth of new furniture. This is treated as a capital expenditure.

- *Room modifications.* Existing classrooms will be modified at a cost of $5,000 each. The amount will be expensed in the first year of operation.

Continuing the worksheet, we add lines for the furniture and the room modifications (Table 20.1).

Other costs can be estimated as follows (Table 20.1):

- *System support costs.* We follow the general guideline, which is that system support costs approximately 30% of equipment cost. We also assume an annual increase in salaries and related benefits of 4%.

- *Maintenance.* We follow the general rule that maintenance approximates 10% of system cost. We further assume an annual

Table 20.1
Distance Education Study—Steps 1–4:
Budgetary Costs

Item	Year 0	Year 1	Year 2	Year 3	Year 4	Year 5
Equipment	$475,000					
Installation		$ 16,000				
Shipping		$ 6,000				
Furniture	$ 30,000					
Modifications		$ 25,000				
Support		$142,500	$148,200	$154,128	$160,293	$166,705
Maintenance		$ 47,500	$ 49,400	$ 51,376	$ 53,431	$ 55,568
Access		$ 5,400	$ 5,400	$ 5,400	$ 5,400	$ 5,400
Usage		$ 20,311	$ 20,311	$ 20,311	$ 20,311	$ 20,311
Total	$505,000	$262,711	$223,311	$231,215	$239,435	$247,984

increase of 10%. This increase is partially due to rising labor costs and other expenses, but it is mostly attributable to the age of the equipment. The longer the equipment is in service, the more frequently it will require service.

Here are some other costs that also need to be included in this worksheet:

- *Network access.* In each site, three ISDN BRI lines will be installed for the exclusive use of the distance education system. The average cost is $360 per year for each line, yielding an annual network access cost of $5,400 (5 sites × 3 lines × $360). These costs have tended to decline in recent years. Therefore, no upward trending is required.

- *Usage.* During the semester, classes meet once a week for 50 minutes for every credit hour. Each link will be initiated 5 minutes before the class begins and left up for 5 minutes after the class ends. Therefore, each credit hour taught at a site equals 1 hour of network usage per week. A semester is 13 weeks long, and there are two semesters a year, making each credit hour of instruction equal to 26 hours of network usage annually.

  Each of the sites will be connected via the network to the MCU. All five sites will be linked for eight 3-credit courses each semester. These classes will result in annual network usage equal to 3,120 hours (5 sites × 8 courses × 3 credits × 26 hours of network usage). Three sites will be linked for six additional 3-credit courses, for additional network usage of 1,404 hours annually (3 sites × 6 courses × 3 credits × 26 hours). Finally, three sites will be linked for two 2-credit courses, adding network usage of 312 hours (3 sites × 2 courses × 2 credits × 26 hours). Total annual usage for these courses is 4,836 hours (3,120 hours + 1,404 hours + 312 hours). The long-distance carrier will charge $0.07 per minute, yielding an annual charge of $20,311 (4,836 hours × 60 minutes × $0.07/minute). These costs are expected to be stable or decline, so no upward trend is applied.

Let's add the network access and usage costs to the worksheet (Table 20.1).

There could be additional costs, such as instructor salaries, associated with providing these classes. To simplify this example, we assume that the course is taught at one site, that the instructor does not receive any additional pay for teaching the remote students simultaneously with the local students, and that no additional expenses were incurred.

We now have a budgetary look at the cost for instituting distance education. We will refer to this later when we perform a discounted cash flow analysis.

20.2.2 Benefits worksheet for implementing distance education

The benefits of implementing visual telephony vary with the application. One likely result of instituting a distance education network is an increase in enrollments. The benefit is new revenue in the form of tuition payments made by students at the remote sites.

Let's continue with our example by making a few assumptions about anticipated revenue:

■ *Revenue.* The university uses the network to provide 92 credit hours of instruction a year (see the usage assumptions given earlier). The average class size, per remote classroom, is eight students. Tuition is $200 per credit hour.

We now construct a simple spreadsheet (Table 20.2) to calculate the additional revenue that would be received if distance education were implemented.

This new revenue can be added to a revenue worksheet (Table 20.3). The calculated amount is entered in Year 1 and trended upward at 4%, since tuition often increases based on increasing faculty salary costs.

20.2.3 Cash flow analysis for implementing distance education

The next worksheet (Table 20.4) combines the costs and benefits of implementing distance education to arrive at the cash flows for the

Table 20.2

Distance Education Study—Step 5: Revenue Opportunity

Courses	Credit Hours (Cr Hr)	Semesters	Annual Credit Hours	Remote Sites	Students per Remote Site	Annual Additional Student Cr Hr	Additional Revenue at $200 per Cr Hr
8	3	2	48	4	8	1,536	$307,200
6	3	2	36	2	8	576	$115,200
2	2	2	8	2	8	128	$ 25,600
Total			92			2,240	$448,000

Table 20.3
Distance Education Study—Step 6:
Revenue Opportunity

Item	Year 0	Year 1	Year 2	Year 3	Year 4	Year 5
Tuition		$448,000	$465,920	$484,557	$503,939	$524,097

Table 20.4
Distance Education Study—Steps 7–8:
Cash Flow Analysis

Item	Year 0	Year 1	Year 2	Year 3	Year 4	Year 5
Costs:						
Equipment	($475,000)					
Installation		($ 16,000)				
Shipping		($ 6,000)				
Furniture	($ 30,000)					
Modifications		($ 25,000)				
Support		($142,500)	($148,200)	($154,128)	($160,293)	($166,705)
Maintenance		($ 47,500)	($ 49,400)	($ 51,376)	($ 53,431)	($ 55,568)
Access		($ 5,400)	($ 5,400)	($ 5,400)	($ 5,400)	($ 5,400)
Usage		($ 20,311)	($ 20,311)	($ 20,311)	($ 20,311)	($ 20,311)
Total	($505,000)	($262,711)	($223,311)	($231,215)	($239,435)	($247,984)
Revenues:						
Tuition		$448,000	$465,920	$484,557	$503,939	$524,097
Total		$448,000	$465,920	$484,557	$503,939	$524,097
Analysis:						
Net cash flow	($505,000)	$185,289	$242,609	$253,342	$264,504	$276,113
Discount factor (Appendix A)	1.0000	0.9524	0.9070	0.8638	0.8227	0.7835
Discounted net cash flow	($505,000)	$176,469	$220,046	$218,837	$217,608	$216,334
Cumulative discounted net cash flow	($505,000)	($328,531)	($108,484)	$110,352	$327,960	$544,294

project for each of the 5 years of its life. Capital costs and expenses represent cash flowing out and are shown as negative numbers (in parentheses). Additional tuition revenues represent cash flowing in and are shown as positive numbers.

Let's further assume that this is a state university. The alternative use of funds is to let the taxpayers keep the money and invest it themselves. A conservative investment could earn 5%, so we will use that figure as our discount rate in Table 20.4.

Sometimes the alternative to a distance education project is to reach remote students in other ways, perhaps by hiring additional instructors and by providing travel expenses to these instructors. In this case, a different set of costs and benefits is incurred. For this example, we assume that the alternative is simply *not* to institute distance education or, in other words, maintain the status quo. In this scenario, we do not incur any costs, nor do we derive the benefit of increased revenue.

Because the financial results of doing nothing are zero, the analysis already completed by itself supports the implementation of distance education using visual telephony. The project payback is 3 years with a net present value at the end of its economic life of $544,294. Using a financial program, the internal rate of return on this project can be determined to be 35.98%.

20.3 A complex study of corporate videoconferencing

Businesses must pay taxes. Tax computations complicate the analysis. The application can also add complexity to an analysis. As an example, this next study assumes a corporation is considering implementing videoconferencing in an effort to reduce travel.

In this analysis, there are two alternatives. "Alternative A" is to install videoconferencing in six locations: at headquarters in Kansas City, Missouri; and at regional offices in Atlanta, Chicago, New York, St. Louis, and San Francisco. "Alternative B" is to continue the business travel required among those cities.

20.3.1 Cost worksheet for implementing videoconferencing

In this business application, costs for equipment, installation, shipping, sales tax, furniture, room modification, maintenance, system support, network access, and usage are treated the same as described for the distance education application. The specific assumptions for each item follow:

- *Equipment cost.* The initial installation will include a custom room at headquarters in Kansas City at a cost of $60,000. It will also include a large group rollabout costing $50,000 in each of five regional offices in Atlanta, Chicago, New York, St. Louis, and San Francisco.

- *Installation, shipping, and sales tax.* Installation charges of $15,000, shipping charges of $6,000, and a 6.4% sales tax on the equipment will be expensed. A cost worksheet would begin like the one shown in Table 20.5.

Cost assumptions for furniture and modifications made to the space assigned for the videoconferencing project are considered next:

- *Furniture.* The furnishings in the custom room at headquarters will cost $20,000. Furniture for the regional offices will cost $5,000 for each room.

- *Room modifications.* The custom room at headquarters will be built in a space that is currently used for offices, incurring costs

Table 20.5

Videoconferencing Study—Steps 1–4:
Budgetary System Costs

Item	Year 0	Year 1	Year 2	Year 3	Year 4	Year 5
Equipment	$310,000					
Installation		$ 15,000				
Shipping		$ 6,000				
Sales tax		$ 19,860				
Furniture	$ 45,000					
Modifications	$120,000	$ 25,000				
Support		$ 93,000	$ 96,720	$100,589	$104,612	$108,797
Maintenance		$ 31,000	$ 34,100	$ 37,510	$ 41,261	$ 45,387
Access		$ 7,400	$ 7,400	$ 7,400	$ 7,400	$ 7,400
Usage		$ 22,500	$ 22,500	$ 22,500	$ 22,500	$ 22,500
Multipoint		$ 6,750	$ 6,750	$ 6,750	$ 6,750	$ 6,750
Total	$475,000	$226,510	$167,470	$174,749	$182,523	$190,834

to modify of $120,000. The regional offices will modify existing conference rooms for $5,000 each.

Costs for furniture and the major construction at headquarters are inserted into the worksheet in Year 0 (Table 20.5). The costs to modify the existing conference rooms in the regional offices are treated as Year 1 expenses.

Next, we consider cost assumptions for maintaining and supporting the videoconferencing systems:

- *System support costs.* We follow the general guideline that assumes system support costs equal 30% of equipment cost. We also assume an annual increase in salaries of 4%.

- *Maintenance.* We use the general rule of 10% of system cost, with an annual increase of 10% attributable to rising labor costs and the aging of the equipment.

These cost assumptions are inserted in the worksheet in Year 1 and trended upward for the remaining years of the system (Table 20.5).

Although there will be a small amount of depreciation not recognized at the end of the study, we limit the analysis to the 5-year economic life of the equipment. The small amount of tax reduction received after 6 years has an insignificant impact on the results of the analysis.

The final cost assumptions concern network access and usage:

- *Network access.* Headquarters has several ISDN PRI access lines, and the videoconferencing system will share some of this capacity. An annual cost of $2,000 is assigned to the videoconferencing application. The regional offices will install three ISDN BRI lines at an average cost of $360 per year per line, totaling $5,400 (5 offices × 3 lines × $360 per line). Total access costs are $7,400 ($2,000 + $5,400). These costs have tended to decline in recent years. Therefore, no trend is applied.

- *Usage.* The corporation estimates that headquarters will be connected in a point-to-point session with each of the five regions an average of one-half hour a day, totaling 625 hours annually, assuming 250 business days per year (5 sites × 0.5 hours × 250 days). There will be point-to-point conferences between regions that add another 2 hours a day, or 500 hours annually (2 hours × 250 days). There also will be a 1-hour multipoint conference connecting all six sites every other day. This adds another 750 hours per year (6 sites × 1 hour × 125 days). Total usage is therefore 1,875 hours annually (625 hours + 500 hours + 750 hours). Network charges are assumed to be $0.20 per channel per minute for a total of $22,500 (1,875 hours × $0.20/min × 60 minutes/hour). In addition, the *multipoint service provider* charges $0.15 per channel per minute, adding $6,750 (750 hours × $0.15 × 60 minutes/hour) to the annual costs. These costs are expected to be stable or decline. No trend is applied.

Note that when two sites are connected in a point-to-point configuration, they share a single duplex connection. When a multipoint service is used, each site has a duplex connection to/from the service provider's MCU. In this example, a six-site multipoint conference requires six connections, whereas a two-site conference requires only a single connection. Let's add these numbers to our worksheet (Table 20.5).

Table 20.5 provides a good budgetary look at the costs of videoconferencing. However, since the organization is a business and subject to taxes, we must convert the budgetary numbers into after tax cash flows. The first step is to recognize the *before tax cash flow*.

Remember that capital and expenses are outflows of cash. Notice in Table 20.6 that these cash flow figures are now in parentheses to reflect a negative amount.

Expenses are tax deductible. Capital outlays are not, but they do provide annual tax write-offs for depreciation. In the next step, we include depreciation to arrive at the taxable income, which is a loss since we are only looking at costs. Then, by multiplying the taxable income by a factor of $1 - t$, where t is the tax rate, we arrive at the *after tax income*.

In Table 20.6, we use a corporate income tax rate of 37%. Note that the capital outlay in Year 0 is not affected by taxes.

Table 20.6

Videoconferencing Study—Steps 5–7:
Cash Flow Analysis of Installing Videoconferencing

Item	Year 0	Year 1	Year 2	Year 3	Year 4	Year 5
Equipment	($310,000)					
Installation		($ 15,000)				
Shipping		($ 6,000)				
Sales tax		($ 19,860)				
Furniture	($ 45,000)					
Modifications	($120,000)	($ 25,000)				
Support		($ 93,000)	($ 96,720)	($100,589)	($104,612)	($108,797)
Maintenance		($ 31,000)	($ 34,100)	($ 37,510)	($ 41,261)	($ 45,387)
Access		($ 7,400)	($ 7,400)	($ 7,400)	($ 7,400)	($ 7,400)
Usage		($ 22,500)	($ 22,500)	($ 22,500)	($ 22,500)	($ 22,500)
Multipoint		($ 6,750)	($ 6,750)	($ 6,750)	($ 6,750)	($ 6,750)
Before tax cash flow	($475,000)	($226,510)	($167,470)	($174,749)	($182,523)	($190,834)
Depreciation		($ 95,000)	($152,000)	($ 91,200)	($ 54,720)	($ 54,720)
Taxable income		($321,510)	($319,470)	($265,949)	($237,243)	($245,554)
After tax income		($202,551)	($201,266)	($167,548)	($149,463)	($154,699)
Depreciation		$ 95,000	$152,000	$ 91,200	$ 54,720	$ 54,720
Net cash flow	($475,000)	($107,551)	($ 49,266)	($ 76,348)	($ 94,743)	($ 99,979)

Since depreciation is not an actual annual outflow of cash, we add it back to get the actual annual cash flow. The reason for including it was simply to enable the calculation of *after tax* income. Finally, in Table 20.6, we recognize the initial outflow of cash for the capital purchases.

Table 20.6 lists the cash flows that result from installing videoconferencing and using it as anticipated. We will use this portion of the analysis again when we do our comparison. Now we must determine the benefits that result from installing videoconferencing.

20.3.2 Benefits that can be derived from videoconferencing

As we discussed in Chapter 1, the benefits of videoconferencing are faster, better decisions; reduced travel costs; greater productivity; and improved employee morale and safety. But how do you quantify a benefit such as faster, better decisions?

Several manufacturing companies have attributed timelier introduction of new products to their extensive use of videoconferencing during the product design process. They could assign the first year's profit from the sale of these products as a benefit of videoconferencing. Having captured market share, they could attribute some profits in Years 2 and 3 to videoconferencing as well.

Another manufacturer documented the discovery and correction of a potential manufacturing defect uncovered during the regular project reviews that videoconferencing made possible.

Yet another manufacturer was able to keep an assembly line operational because a headquarters engineer was able to diagnose a problem with the equipment and talk the local technicians through repairing it.

Unfortunately, these companies, and most others that have instituted videoconferencing, recognize these benefits only *after* installing the videoconferencing network and integrating it into their corporate culture. As one vice-president of engineering put it, ''We don't plan to build products with defects.''

As a result, these benefits are usually not quantified and included in financial analyses prepared to justify the purchase of a system. Instead, most organizations justify videoconferencing on the reduction of travel costs alone, which are much easier to anticipate and quantify. We will do that first in this example. Then we will add some other benefits and repeat the analysis.

20.3.3 Travel cost worksheet for implementing videoconferencing

The company headquarters is in Kansas City, Missouri, a large city in the center of the United States. Regional offices are in Atlanta, Chicago, New York, St. Louis, and San Francisco. The matrix shown in Table 20.7 reflects the anticipated trips avoided, based on the videoconferencing usage assumptions described earlier and on typical travel patterns in corporations.

We are assuming there that the point-to-point videoconferences between headquarters and the regions replace one trip a week that headquarters personnel take to the regions. Two people usually travel to attend these meetings (2 people × 52 weeks = 104 person trips). The point-to point-videoconferences between regions replace one trip

Table 20.7
Videoconferencing Study—Step 8: Person Trips Avoided

Cities	Kansas City	Atlanta	Chicago	New York	St. Louis	San Francisco
Kansas City		104	104	104	104	104
Atlanta	24		6	6	6	6
Chicago	24	6		6	6	6
New York	24	6	6		6	6
St. Louis	24	6	6	6		6
San Francisco	24	6	6	6	6	

per month between regional offices. Normally one person travels each month to attend these meetings; the personnel alternate traveling. Therefore, one trip from each site to each site is avoided every other month (1 person × 6 months = 6 person trips). Finally, the multipoint conferences are expected to replace two trips per month by one person from each region traveling to headquarters (1 person × 2 trips × 12 months = 24 person trips). We plot the travel that could be avoided on the matrix of Table 20.7.

It is tempting to treat a trip from New York to Kansas City and a trip from Kansas City to New York as the same. While the airfare is the same, local transportation, parking, hotels, and meals cost more in New York. It is important, therefore, to obtain cost information for *each* travel city.

With information from travel agents and from expense reports filed by past travelers, in Table 20.8 we calculate the average cost of one person traveling between the various city pairs.

Then, by combining the cost elements, in Table 20.9 we arrive at an estimate of the cost of each trip.

Using this trip cost information and the previous assumptions about who typically travels, we now calculate in Table 20.10 an estimated annual savings from avoiding travel to conduct selected meetings.

Next, we complete the discounted cash flow analysis. We begin by determining the after tax cash flows from travel that videoconferencing can eliminate (Table 20.11). Because travel costs are highly comprised of labor, we will trend these costs to increase 2% annually.

Table 20.8

Videoconferencing Study—Step 9: Travel Cost Elements

City Pair	To/From Airport	Parking	Airfare	Car Rental/ Taxi Rental	Hotel	Meals
Atlanta → Kansas City	$28	$16	$ 510	$55	n/a	$ 40
Chicago → Kansas City	$24	$20	$ 335	$55	n/a	$ 40
New York → Kansas City	$25	$24	$ 765	$55	n/a	$ 40
St. Louis → Kansas City	$30	$12	$ 160	$55	n/a	$ 40
San Francisco → Kansas City	$28	$30	$1,150	$55	$ 75	$ 80
Kansas City → Atlanta	$30	$12	$ 510	$60	n/a	$ 50
Kansas City → Chicago	$30	$12	$ 335	$60	n/a	$ 55
Kansas City → New York	$30	$12	$ 765	$40	n/a	$ 70
Kansas City → St. Louis	$30	$12	$ 160	$40	n/a	$ 40
Kansas City → San Francisco	$30	$24	$1,150	$60	$100	$130
Atlanta → Chicago	$28	$16	$ 440	$60	n/a	$ 55
Atlanta → New York	$28	$16	$ 530	$40	n/a	$ 70
Atlanta → St. Louis	$28	$16	$ 350	$40	n/a	$ 40
Atlanta → San Francisco	$28	$32	$1,540	$60	$100	$130
New York → Atlanta	$25	$24	$ 530	$60	n/a	$ 50
New York → Chicago	$25	$24	$ 500	$60	n/a	$ 55
New York → St. Louis	$25	$24	$ 605	$40	n/a	$ 40
New York → San Francisco	$25	$48	$1,820	$60	$100	$130
Chicago → Atlanta	$24	$20	$ 440	$60	n/a	$ 50
Chicago → New York	$24	$20	$ 500	$40	n/a	$ 70
Chicago → St. Louis	$24	$20	$ 410	$40	n/a	$ 40
Chicago → San Francisco	$24	$40	$1,320	$60	$100	$130
St. Louis → Atlanta	$30	$12	$ 350	$60	n/a	$ 50
St. Louis → Chicago	$30	$12	$ 410	$60	n/a	$ 55
St. Louis → New York	$30	$12	$ 605	$40	n/a	$ 70
St. Louis → San Francisco	$30	$24	$1,310	$60	$100	$130
San Francisco → Atlanta	$28	$30	$1,540	$60	$ 75	$ 50
San Francisco → Chicago	$28	$30	$1,320	$60	$100	$ 55
San Francisco → New York	$28	$30	$1,820	$40	$125	$140
San Francisco → St. Louis	$28	$30	$1,310	$40	$ 85	$ 40

These travel costs represent before tax cash flows. Because these expenses are outflows of cash, they are represented by negative numbers in the next analysis. In Table 20.12 we show the travel savings as before tax cash flows, all of which would be included in the corporation's taxable income. Next, we calculate the after tax income, based

Table 20.9

Videoconferencing Study—Step 10: Typical Travel Costs

City Pair	Travel Cost
Atlanta → Kansas City	$ 649
Chicago → Kansas City	$ 474
New York → Kansas City	$ 909
St. Louis → Kansas City	$ 297
San Francisco → Kansas City	$1,418
Kansas City → Atlanta	$ 662
Kansas City → Chicago	$ 492
Kansas City → New York	$ 917
Kansas City → St. Louis	$ 282
Kansas City → San Francisco	$1,494
Atlanta → Chicago	$ 599
Atlanta → New York	$ 684
Atlanta → St. Louis	$ 474
Atlanta → San Francisco	$1,890
New York → Atlanta	$ 689
New York → Chicago	$ 664
New York → St. Louis	$ 734
New York → San Francisco	$2,183
Chicago → Atlanta	$ 594
Chicago → New York	$ 654
Chicago → St. Louis	$ 534
Chicago → San Francisco	$1,674
St. Louis → Atlanta	$ 502
St. Louis → Chicago	$ 567
St. Louis → New York	$ 757
St. Louis → San Francisco	$1,654
San Francisco → Atlanta	$1,783
San Francisco → Chicago	$1,593
San Francisco → New York	$2,183
San Francisco → St. Louis	$1,533

on the corporation's assumed tax rate of 37%. This gives us the net cash flow of the travel option.

We can now compare the costs of installing videoconferencing with the benefits, namely, the avoidance of travel expenses. To do this, we create a comparison worksheet (Table 20.13) of the net cash flows for installing videoconferencing and the net cash flow of continuing to travel. Refer to the previous worksheets and use the net cash

Table 20.10

Videoconferencing Study—Step 11: Annual Savings From Travel Avoidance

City Pair	Person Trips	Per Trip Cost	Annual Travel Cost
Atlanta → Kansas City	24	$ 649	$ 15,576
Chicago → Kansas City	24	$ 474	$ 11,376
New York → Kansas City	24	$ 909	$ 21,816
St. Louis → Kansas City	24	$ 297	$ 7,128
San Francisco → Kansas City	24	$1,418	$ 34,032
Kansas City → Atlanta	104	$ 662	$ 68,848
Kansas City → Chicago	104	$ 492	$ 51,168
Kansas City → New York	104	$ 917	$ 95,368
Kansas City → St. Louis	104	$ 282	$ 29,328
Kansas City → San Francisco	104	$1,494	$155,376
Atlanta → Chicago	6	$ 599	$ 3,594
Atlanta → New York	6	$ 684	$ 4,104
Atlanta → St. Louis	6	$ 474	$ 2,844
Atlanta → San Francisco	6	$1,890	$ 11,340
New York → Atlanta	6	$ 689	$ 4,134
New York → Chicago	6	$ 664	$ 3,984
New York → St. Louis	6	$ 734	$ 4,404
New York → San Francisco	6	$2,183	$ 13,098
Chicago → Atlanta	6	$ 594	$ 3,564
Chicago → New York	6	$ 654	$ 3,924
Chicago → St. Louis	6	$ 534	$ 3,204
Chicago → San Francisco	6	$1,674	$ 10,044
St. Louis → Atlanta	6	$ 502	$ 3,012
St. Louis → Chicago	6	$ 567	$ 3,402
St. Louis → New York	6	$ 757	$ 4,542
St. Louis → San Francisco	6	$1,654	$ 9,924
San Francisco → Atlanta	6	$1,783	$ 10,698
San Francisco → Chicago	6	$1,593	$ 9,558
San Francisco → New York	6	$2,183	$ 13,098
San Francisco → St. Louis	6	$1,533	$ 9,198
Total			$621,686

flow entries for videoconferencing (Table 20.6, Step 7, last line) and for travel (Table 20.12, Step 13, last line). Remember that the word *versus* in this comparison can be read as *minus*. To subtract a negative number, we make the number positive and add it.

Now we must do a discounted cash flow analysis to get the complete picture (Table 20.14). We apply the weighted average cost of capital

Table 20.11
Videoconferencing Study—Step 12:
Cash Flow Analysis of Avoidable Travel

Item	Year 0	Year 1	Year 2	Year 3	Year 4	Year 5
Travel savings		$621,686	$634,120	$646,802	$659,738	$672,933

Table 20.12
Videoconferencing Study—Step 13:
Cash Flow Analysis of Avoidable Travel

Item	Year 0	Year 1	Year 2	Year 3	Year 4	Year 5
Before tax cash flow		($621,686)	($634,120)	($646,802)	($659,738)	($672,933)
Taxable income		($621,686)	($634,120)	($646,802)	($659,738)	($672,933)
After tax income		($391,662)	($399,495)	($407,485)	($415,635)	($423,948)
Net cash flow		($391,662)	($399,495)	($407,485)	($415,635)	($423,948)

Table 20.13
Videoconferencing Study—Step 14:
Cash Flow Analysis of Alternatives

Item	Year 0	Year 1	Year 2	Year 3	Year 4	Year 5
Net cash flow, video-conferencing alternative (Step 7)	($475,000)	($107,551)	($ 49,266)	($ 76,348)	($ 94,743)	($ 99,979)
Net cash flow, travel alternative (Step 13)		($391,662)	($399,495)	($407,485)	($415,635)	($423,948)
Net cash flow, comparison of alternatives	($475,000)	$284,111	$350,229	$331,138	$320,892	$323,969

Table 20.14
Videoconferencing Study—Step 15: Cash Flow Analysis of Alternatives

Item	Year 0	Year 1	Year 2	Year 3	Year 4	Year 5
Net cash flow, video-conferencing alternative	($475,000)	($107,551)	($ 49,266)	($ 76,348)	($ 94,743)	($ 99,979)
Net cash flow, travel alternative		($391,662)	($399,495)	($407,485)	($415,635)	($423,948)
Net cash flow, comparison of alternatives	($475,000)	$284,111	$350,229	$331,138	$320,892	$323,969
Discount factor (Appendix A)	1.0000	0.9091	0.8264	0.7513	0.6830	0.6209
Discounted net cash flows	($475,000)	$258,285	$289,430	$248,784	$219,169	$201,152
Cumulative discounted cash flows	($475,000)	($216,715)	$ 72,715	$321,498	$540,667	$741,820

to arrive at the discounted net cash flows. The corporation has a weighted average cost of capital for projects of this type of 10%. Use the discount table given in Appendix A. To complete the analysis, we sum the discounted net cash flows to determine the cumulative discounted cash flows.

This undertaking has a net present value of $741,820. The discounted payback period is 2 years. The internal rate of return is 60%.

20.3.4 Accounting for other benefits

Despite the detail required, travel savings is the easiest benefit to quantify. However, as we noted earlier, the more significant benefits of videoconferencing are derived from the productivity enhancements made possible by corporate use of the system. For example, some organizations recognize increased revenue from getting products to market before their competitors. The additional revenue from quicker product development is not all profit. There are costs associated with making and selling the product. If we use the organization's profit margin, we can determine the increased profit from additional revenues.

- *Revenue increases.* The corporation anticipates that its use of videoconferencing will enable it to bring a product to market a year in advance of its normal target introduction, adding $1,000,000 in revenues. In the following 3 years, a portion of the product's revenue is attributable to its early introduction.

- *Margin.* The profit margin for this product is 10%.

Plugging the numbers into our worksheet, we have the results shown in Table 20.15.

Other productivity improvements reduce costs. For example, if videoconferencing improved employee morale, there would be a corresponding savings in human resources (HR) expenses from a reduction in the number of new employees that must be recruited, hired, moved, and trained. For our example, let's assume that the HR department estimates savings of $33,000 annually as a result of using videoconferencing.

The next worksheet (Table 20.16) includes videoconferencing costs, travel savings, and HR savings. Rather than structure the analysis as a comparison, this time we show the anticipated financial results on a layout similar to an income statement. We can include a discounted cash flow analysis on the same form. Note that the expense entries for travel and HR are positive numbers to show a *savings* in expenses rather than an additional outlay.

Table 20.15

Videoconferencing Study—Step 16:
Increased Profits Attributable to Videoconferencing

Item	Year 0	Year 1	Year 2	Year 3	Year 4	Year 5
New revenue		$0	$1,000,000	$750,000	$500,000	$250,000
Profit margin		10%	10%	10%	10%	10%
Increased profits		$0	$ 100,000	$ 75,000	$ 50,000	$ 25,000

Table 20.16
Videoconferencing Study—Step 17:
Analysis of Videoconferencing Implementation Plan

Item	Year 0	Year 1	Year 2	Year 3	Year 4	Year 5
Revenues:						
New profits		$ 0	$100,000	$ 75,000	$ 50,000	$ 25,000
Expenses:						
Installation		($ 15,000)				
Shipping		($ 6,000)				
Sales tax		($ 19,860)				
Modifications		($ 25,000)				
Support		($ 93,000)	($ 96,720)	($100,589)	($104,612)	($108,797)
Maintenance		($ 31,000)	($ 34,100)	($ 37,510)	($ 41,261)	($ 45,387)
Access		($ 7,400)	($ 7,400)	($ 7,400)	($ 7,400)	($ 7,400)
Usage		($ 22,500)	($ 22,500)	($ 22,500)	($ 22,500)	($ 22,500)
Multipoint		($ 6,750)	($ 6,750)	($ 6,750)	($ 6,750)	($ 6,750)
Depreciation		($ 95,000)	($152,000)	($ 91,200)	($ 54,720)	($ 54,720)
Travel savings		$621,686	$634,120	$646,802	$659,738	$672,933
HR savings		$ 33,000	$ 33,000	$ 33,000	$ 33,000	$ 33,000
Expenses		$333,176	$347,650	$413,853	$455,495	$460,379
Analysis:						
Net taxable income		$333,176	$447,650	$488,853	$505,495	$485,379
Net after tax income		$209,901	$282,019	$307,978	$318,462	$305,789
Capital	($475,000)					
Depreciation		$ 95,000	$152,000	$ 91,200	$ 54,720	$ 54,720
Net cash flow	($475,000)	$304,901	$434,019	$399,178	$373,182	$360,509
Discount factor (Appendix A)	1.0000	0.9091	0.8264	0.7513	0.6830	0.6209
Discounted net cash flows	($475,000)	$277,185	$358,674	$299,902	$254,883	$223,840
Cumulative discounted net cash flows	($475,000)	($197,815)	$160,859	$460,761	$715,644	$939,484

As Table 20.16 shows, this undertaking has a net present value of $939,484. The discounted payback period is 2 years. The internal rate of return is 71%.

20.4 Summary

The examples in this chapter should help identify the information that should be gathered to analyze a proposed implementation of visual telephony. As the costs associated with visual telephony continue to decline and as global competition intensifies, and as travel costs climb, the wisdom of implementing visual telephony will be more obvious intuitively. Until then, it may be necessary to perform an analysis similar to the samples provided in this chapter.

Good luck!

Part 6

Appendixes

H ERE IS SOME additional material for reference purposes. Appendix A is the discount table. Appendix B is a list of IRS-allowed depreciation rates. Appendix C is a short description of how to calculate monitor size. Appendix D contains a glossary of technical and financial terms used in this book.

Discount Factors

Table A.1
Discount Factors—10% Through 15%

Year	10%	11%	12%	13%	14%	15%
1	0.9091	0.9009	0.8929	0.8850	0.8772	0.8696
2	0.8264	0.8116	0.7072	0.7831	0.7695	0.7561
3	0.7513	0.7312	0.7118	0.6931	0.6750	0.6575
4	0.6830	0.6587	0.6355	0.6133	0.5921	0.5718
5	0.6209	0.5935	0.5674	0.5428	0.5194	0.4972

Table A.2
Discount Factors—16% Through 25%

Year	16%	17%	18%	19%	20%	25%
1	0.8621	0.8547	0.8475	0.8403	0.8333	0.8000
2	0.7432	0.7305	0.7182	0.7062	0.6944	0.6400
3	0.6407	0.6244	0.6086	0.5934	0.5787	0.5120
4	0.5523	0.5337	0.5158	0.4987	0.4823	0.4096
5	0.4761	0.4561	0.4371	0.4190	0.4019	0.3277

B

Depreciation Factors for Tax Purposes

Table B.1
Depreciation Factors

Period	Factor
Year 1	0.2000
Year 2	0.3200
Year 3	0.1920
Year 4	0.1152
Year 5	0.1152
Year 6	0.0576

C

Calculating Monitor Size and Viewing Distance

THIS INFORMATION SUPPLEMENTS *the discussion on monitor size contained in Chapter 9. See also Chapter 13 for information on room dimensions.*

The correct monitor size can be calculated based on the number of participants that will be viewing the display.

Actually, the calculations begin with determining the distance the participants must be seated from the *camera* to allow all the active participant seating positions to be imaged. Because the camera must be located near the center of the viewing surface, this is also the distance the participants will be from the monitor. Once this distance is established, the size of the monitor can be calculated.

Chapter 13 contains a detailed explanation of how the various dimensions of a video telephony installation are calculated. As described there, the typical conference room chair is 2 feet wide. Therefore, to image one conference participant the camera will need to capture 2 feet. If more than one person is in a view, there must be at least 1 foot between their chairs. The camera will need to image at least 5 feet when there are two participants and 8 feet for three participants (Table C.1).

Table C.1
Imaging Width

Number of Participants	Width of Picture (feet)
1	2
2	5
3	8
4	11

The typical camera lens used in videoconferencing can image a 60-degree area. Using simple geometry, in Table C.2 we determine the acceptable distances from the camera. (Use the formula for the circumference of a circle, $\pi 2r$. A 60-degree arc is 1/6 of the total circumference. Therefore, six times the width of the picture divided by π divided by 2 will give the radius of the circle, or the distance the camera should be from the participants.)

Table C.2
Distance From Camera Calculations

Number of Participants	Width of Picture (feet)	Distance From Camera (feet)
1	2	2.0
2	5	5.0
3	8	8.0
4	11	10.5

Because the camera is placed near the display, the distance an observer is from the screen is approximately the same as the distance from the camera.

Ideally, a person should be 7.5 to 8.0 times the distance from a monitor as the height of the monitor. Now that we know the distance from the display, we can calculate the height of the display by dividing the distance (in inches) by a factor of 7.5 or 8. But monitors are not provided based on their height, but rather on a diagonal measure in inches. Since we know the height and the fact that the width is 4/3 of the height (because of the 4:3 aspect ratio of video displays), we can calculate the diagonal using the formula $a^2 + b^2 = c^2$. Using these formulas, we determine the *minimum* monitor sizes, measured diagonally in inches, as shown in Table C.3.

Table C.3

Minimum Monitor Sizes

Number of Participants	Distance From Display (feet)	Screen Size (inches)
1	2.0	5
2	5.0	13
3	8.0	21
4	10.5	27

D

Glossary

Access line	Telephone line connecting the user directly to a long distance carrier.
ADPCM	Abbreviation for adaptive differential pulse code modulation. A method of coding voice frequencies into a digital format.
Algorithm	Sequence of mathematical processes.
Analog transmission	A method of sending signals—voice, video, data—in which the transmitted signal is analogous to the original signal. A continuous electical wave of varying intensity and frequency is used to send the information.
Arraignment	Formal reading, before a judge, of charges against a prisoner.

Asynchronous transfer mode (ATM)	A form of data transmission. Information is placed in packets that are a fixed size. The data from an individual user may be sent as required rather than at fixed times. In that sense, the transmission is asynchronous.
ATM	Asynchronous transfer mode.
Bandwidth	The effective carrying capacity of a telecommunications transmission facility.
Basic rate interface (BRI)	Access line to the ISDN. It is used to connect desktop terminals and telephones. Each BRI line has two B channels of 64 kbps each and one D channel of 16 kbps.
B channel	Bearer channel. A time interval during which 64 kbps of information is sent over a digital transmission facility.
Bit	Binary digit. The smallest unit of information a computer recognizes. It can either be a mark (pulse of current) or space (absence of current).
Bonding	From bandwidth on demand. Method used to synchronize data channels.
BRI	Basic rate interface.
Bridge	A device used to connect several users on a single multi-party conference call.
Bypass	Telephone line provided by a long distance carrier that does not have access to a local exchange carrier's network.
Capital	Money used to purchase an asset with a useful life of more than one year.
Cascading	Method of tying MCUs together.
Cash Flow	Money that actually enters and leaves an organization.
CCD	Charged coupled device. Solid state device used to convert light into electronic signals. Used in most video cameras.

CCITT	International Telegraph and Telephone Consultative Committee. The original name of the ITU.
Channel	Physical path for analog or digital transmission of voice, data, facsimile, or video between two terminals or telephones.
Channel service unit (CSU)	A device used to interface between digital customer equipment and the telephone company demarcation point.
CIF	Common Intermediate Format. The video format selected for international compressed video telephony. It shares characteristics of both American and European video formats. The resolution is 354×288.
Codec	An acronym for coder-decoder. A device that converts analog signals into digital format and digital bit streams into analog signals.
Compression	Any of several processes used to reduce the number of bits required to represent information that is stored or transmitted.
Continuous presence	In videoconferencing, the term is used to describe the transmission of two half pictures as one image. In multipoint conferencing, the term is used to describe the display of all conference participants in windows on a single video screen.
Cost benefit analysis	Study of all costs and benefits pertinent to a decision.
CSU	Channel service unit.
Data	Information represented in digital form.
D channel	Dialing channel. The portion of an ISDN transmission used for call set-up information.
Deposition	Statement given under oath.
Depreciation	An expense equal to a portion of an asset's value.
DES	Data Encryption Standard. Method of encrypting digital transmissions approved by the National Security Agency.

DS-0	A 64 kbps channel.
DS-1	The Level 1 standard for digital systems, consisting of 24 DS-0 channels. Also known as T1.
DS-3	The Level 3 standard for digital systems consisting of 45 Mbps. Also known as T3.
DSU	Digital service unit. A device used to interconnect digital equipment to a digital telephone line. Used in conjunction with a CSU. It is used to convert the user data into bipolar format.
Dedicated network	A network made up of transmission facilities that are exclusively used by one customer.
Digital transmission	A method of sending information coded with on and off pulses of electricity.
Discount rate	The interest rate an organization can expect to earn. Used to determine the present value of money.
Duplex	Bi-directional. The term is most often used to describe voice connections that allow the users to speak and hear at the same time.
Echo cancellation	A device that eliminates echo by isolating and filtering out the return of a received signal.
FCIF	Full CIF. Same as CIF.
Fractional T1	Phone company offering that is based on T1 access line, with a customer allotted a portion for transmission.
Frame	A complete screen of video.
fps	Frames per second. The number of complete screens of video that is transmitted every second.
Future value	What a sum of money today would be worth in the future if invested at the discount rate.
H.221	Call management portion of H.320.
H.243	Multipoint chairman control under H.320.
H.261	Video coding standard under H.320.
H.263	Alternative video coding standard under H.320.
H.281	Far end camera control standard under H.320.

H.320	ITU international standard for visual telephony.
H channel	High capacity digital channel.
Half duplex	Voice transmissions that are capable of transmitting or receiving, but not both at the same time.
Hertz (Hz)	The number of cycles per second of a frequency.
Hz	Abbreviation for hertz.
Inband signaling	Term used to describe circuits where call set-up information is carried within the circuit.
Inter-LATA	Telephone traffic between LATAs.
Internal rate of return	The interest rate an investment earns.
Interoperability	Compatibility between equipment.
ISDN	Integrated services digital network. International public switched digital network.
ITU	International Telecommunications Union. United Nations organization with responsibility for international telephone standards.
JPEG	Joint Photographic Experts Group. Standard for encoding still images in computer files.
kbps	Kilobits per second. One thousand bits per second.
kHz	Kilohertz. One thousand hertz.
LAN	Local area network. Private data network of PCs, peripherals, and other equipment linked to a local packet switch, usually serving a cluster of geographically related offices or buildings.
LATA	Local access transport area. Local area in which there is a concentration of telephone traffic.
LEC	Local exchange carrier. Telephone company providing local telephone service.
MACRS	Modified Asset Cost Recovery System. The IRS-approved method of accelerated depreciation.
Mark	One bit of digital information consisting of a pulse of current. Also called a *one*.

Mbps	Megabits per second. One million bits per second.
MCU	Multipoint control unit.
Multiplexer (mux)	A device for combining two or more signals for transmission over the same transmission path.
Multipoint control unit (MCU)	A device used to combine point-to-point visual telephony calls into one multi-location call. Similar to a voice bridge.
Mux	Multiplexer.
Net present value (NPV)	The algebraic sum of all present values in a study.
Network	A number of communications paths that allow information to flow between user equipment.
NT1	Network termination device required on BRI access lines.
NTSC	Television standard derived from the National Television Standards Committee. The organization that adopted the standard in 1954. This standard is used throughout North America, Japan, and elsewhere. It consists 30 frames per second of video at 525 × 480 resolution lines.
Packet	A unit of digital data with a set number of bits, including some bits that serve as an address code. Packets can be sent through a packet switched network by the best route, and reunited at the destination, regardless of the order in which they arrive.
PAL	Phased Alternating Line. European television standard consisting of 50 frames per second of video at 625 × 576 resolution.
Payback	The amount of time an investment takes to return its original cost.
PBX	Private branch exchange. A switchboard for routing calls within a facility.
PCM	Pulse code modulation. A method of coding voice frequencies into a digital format.

Pel	Picture element.
PIC	Primary interexchange carrier. The long distance carrier selected by a customer.
Picture element	Basic unit of video information. Also known as *pixel* and *pel*.
Pixel	Picture element.
Port	Physical interface on a computer, used by a remote terminal to gain access to the computer's processor and memory.
Present value (PV)	What a sum of money to be received in the future is worth if received today.
PRI	Primary rate interface.
Primary data rate	1.544 Mbps in the United States and Japan, 2.048 Mbps in Europe and other parts of Asia.
Primary rate interface (PRI)	In the United States and Japan, 23 B channels and one D channel. In Europe, 30 B channels and one D channel. All channels are 64 kbps.
Protocol	A set of rules which define procedures for the transfer of information in a computing or communications system.
PSTN	Public switched telephone network. The combined transmission facilities of the world's telephone companies and administrations, including all circuits available to users on an unrestricted basis.
QCIF	Quarter Common Intermediate Format. Video resolution defined in ITU recommendation H.320, equal to 176 × 188. See also CIF.
Refresh rate	The number of frames of video per second transmitted.
Rollabout system	Factory-built videoconferencing system that can be quickly installed in a conference room. Most systems have casters and can be rolled into place.
SECAM	Sequentiel Couleur Avec Memoire, or Sequential Color with Memory. French television standard consisting of 50 frames per second of video at 625 × 576 resolution.

SONET	Synchronous optical network. A standard for fiber optics as the transport for information transfer.
Space	One bit of digital information consisting of no current. Also called a *zero*.
Split screen	*See* Continuous **presence**.
Subscriber line	Telephone line from user's home or office to the nearest telephone company office.
System integrator	Organization that combines components into a complete, coherent system.
SVGA	Super visual graphics array. Computer monitor resolution of 1,280 × 1,024.
T1	Popular name given to DS-1 service. The data rate is 1.544 Mbps.
T.120	ITU standard for data transfer.
Tandem switch	Telephone switch used to interconnect other switches.
Telemedicine	The use of visual telephony for remote diagnostics, reading X-rays, and remote case management.
Teleradiology	Transmission of digital images of X-ray films for remote assessment.
TDM	Time division multiplexing. Method of mixing signals by interleaving them in time.
Trunk	The connection between telephone switches.
VGA	Visual graphics array. Computer monitor resolution of 1,024 × 768.
Videoconferencing	The use of visual telephony to replace an in-person meeting.
Visual telephony	The transmission of video over telephone lines.

About the Authors

Edward A. Daly has more than 25 years of experience in the telecommunications field, a majority of which has been in visual telephony.

He began as a financial analyst and manager of economic studies for AT&T. In succeeding years, he held sales and marketing positions with AT&T, GPT Video Systems, Panasonic Broadcast Television Systems, BT, and Tandberg.

Mr. Daly graduated cum laude from Herbert H. Lehman College (City College of New York) with a B.A. in economics and received an M.B.A. from Fordham University.

He is the father of three grown sons and lives with his wife in Ridgewood, New Jersey.

Kathleen J. Hansell has had dual careers in education and telecommunications. She has been a college professor of business and has been the lead researcher and consultant in a telecommunications research business that specialized in video communications.

Dr. Hansell holds an Ed.B. in business education, an Ed.M. in instructional communications, and an Ed.D. in curriculum planning and research from State University of New York at Buffalo.

She lives in Atlanta, Georgia, with her husband and two children.

Index

Computer Mediated Communications: Multimedia Applications, Rob Walters

Computer Telephony Integration, Second Edition, Rob Walters

Convolutional Coding: Fundamentals and Applications, Charles Lee

Desktop Encyclopedia of the Internet, Nathan J. Muller

Digital Beamforming in Wireless Communications, John Litva and Titus Kwok-Yeung Lo

Distributed Multimedia Through Broadband Communications Services, Daniel Minoli and Robert Keinath

Distance Learning Technology and Applications, Daniel Minoli

Electronic Mail, Jacob Palme

Enterprise Networking: Fractional T1 to SONET, Frame Relay to BISDN, Daniel Minoli

FAX: Digital Facsimile Technology and Applications, Second Edition, Dennis Bodson, Kenneth McConnell, and Richard Schaphorst

FDDI and FDDI-II: Architecture, Protocols, and Performance, Bernhard Albert and Anura P. Jayasumana

Guide to Telecommunications Transmission Systems, Anton A. Huurdeman

A Guide to the TCP/IP Protocol Suite, Floyd Wilder

Implementing X.400 and X.500: The PP and QUIPU Systems, Steve Kille

Information Superhighways Revisited: The Economics of Multimedia, Bruce Egan

International Telecommunications Management, Bruce R. Elbert

International Telecommunication Standards Organizations, Andrew Macpherson

Radio-Relay Systems, Anton A. Huurdeman

Signaling in ATM Networks, Raif O. Onvural, Rao Cherukuri

Smart Cards, José Manuel Otón and José Luis Zoreda

Smart Card Security and Applications, Mike Hendry

SNMP-Based ATM Network Management, Heng Pan

Successful Business Strategies Using Telecommunications Services, Martin F. Bartholomew

Super-High-Definition Images: Beyond HDTV, Naohisa Ohta, et al.

Telecommunications Deregulation, James Shaw

Telemetry Systems Design, Frank Carden

Teletraffic Technologies in ATM Networks, Hiroshi Saito

Understanding Modern Telecommunications and the Information Superhighway, John G. Nellist and Elliott M. Gilbert

Visual Telephony, Edward A. Daly and Kathleen J. Hansell

World-Class Telecommunications Service Development, Ellen P. Ward

For further information on these and other Artech House titles, including previously considered out-of-print books now available through our In-Print-Forever™ (IPF™) program, contact:

Artech House
685 Canton Street
Norwood, MA 02062
781-769-9750
Fax: 781-769-6334
email: artech@artech-house.com

Artech House
46 Gillingham Street
London SW1V 1AH England
+44 (0) 171-973-8077
Fax: +44 (0) 171-630-0166
email: artech-uk@artech-house.com

Find us on the World Wide Web at:
www.artechhouse.com